Bernard J. Cigrand

The Rise, Fall and Revival of Dental Prosthesis

Bernard J. Cigrand

The Rise, Fall and Revival of Dental Prosthesis

ISBN/EAN: 9783337426682

Printed in Europe, USA, Canada, Australia, Japan

Cover: Foto ©berggeist007 / pixelio.de

More available books at **www.hansebooks.com**

Dr. JOSEPH LEMAIRE,

(Surgeon-Dentist,)

FIRST PRACTICING DENTIST IN AMERICA.

Seal of the Congress.

ART INSTITUTE, CHICAGO.

MEETING PLACE OF COLUMBIAN DENTAL CONGRESS.

THE

RISE, FALL AND REVIVAL

OF

DENTAL PROSTHESIS

BY

B. J. CIGRAND, B. S., D. D, S.,

PROFESSOR OF DENTAL PROSTHESIS AND METALLURGY IN THE
AMERICAN COLLEGE OF DENTAL SURGERY; LECTURER IN THE
PRACTITIONER'S POST-GRADUATE COURSE OF THE SAME INSTI-
TUTION; MEMBER OF THE ILLINOIS STATE DENTAL SO-
CIETY; THE CHICAGO DENTAL SOCIETY; THE CHICAGO
DENTAL CLUB; THE HAYDEN DENTAL SOCIETY;
THE DENTAL LEGAL ASSOCIATION OF ILLI-
NOIS; THE COLUMBIAN DENTAL CLUB, AND
KINDRED DENTAL SOCIETIES.

———

FULLY ANNOTATED AND ILLUSTRATED

———

Second Edition, Revised and Enlarged

———

CHICAGO.
THE PERIODICAL PUBLISHING CO.,
434 WABANSIA AVENUE.

To HENRY J. M'KELLOPS, D. D. S ,
Who has ardently labored in behalf of the dental
profession in gathering home the published
records extant on dental subjects; and has
thus created and established the largest
and most complete dental bibliotheke
in the world. In these "Archives of
Dental Literature" the present
volume was evolved, and for
the liberal hospitality shown
the writer a tribute of
gratefulness is mani-
fested in this in-
scription by the
author.

CONTENTS.

PLATE I.

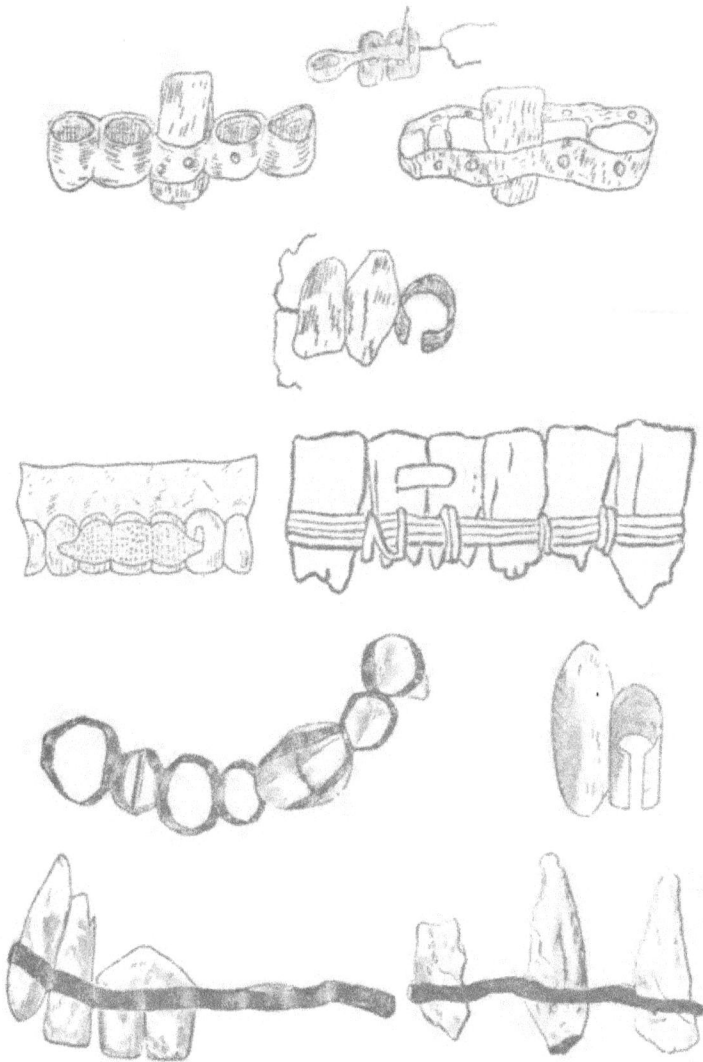

Specimens of Ancient Dental Art

PREFACE TO FIRST EDITION.

PROEM TO STUDENTS.

The title of this lecture is so fully indicative of its character that scarcely a word of "foretalk," as the Saxons express it, seems necessary. As this is my initiatory talk to you, my lecture* will be of a prefatory character in inaugurating the course in Dental Prosthesis. All that is requisite by way of preface here, is to give brief account of the within contained remarks. This proem may, at the same time, serve as an apology for some of the defects of which the author is fully conscious.

Believing it necessary that the dental student, in beginning the study of his chosen profession, should know something of its antique birth, growth and development, I felt it a duty incumbent on myself to open our acquaintance by cordially introducing to you the history of this branch of dental art and science.

---

*The first edition was the outcome of four lectures delivered before the students of the American College of Dental Surgery, 1892.

Knowing full well that many of you before coming here had little knowledge relative to this interesting theme, on account of non-access to journals, I am confident we can profitably spend our first hours.

In many of our dental schools, during the entire three or more years that the student serves within their walls, not one sentence does he hear concerning the struggles and advancements of his prehistoric-professional-forefathers; and, sorry to say, the worthy college from which I hail was addicted to this apparent neglect.

Many months have been spent in gleaning the contents of this lecture, which is the result, rather, of occasional moments of leisure from the duties of an active professional life than of a special devotion to the cultivation of any superb thesis on the subject.

Such sources of information were consulted as were deemed advantageous towards compiling and completing the lecture. Among these "help-mates" I gladly mention: Cosmos, Review, Items of Interest and many voluminous references in the various libraries.

In conclusion, will add, should this published research merit a dedicatorial note, I am certain none more worthily deserve the inscription than my many kind and thorough instructors, and more especially Drs. Haskell and Sherwood, whose untiring devotion has placed them in my memory's safe-keeping.

Very Sincerely,

D. J. Cigrand.

CHICAGO, Sept. 21, 1892.

PREFACE TO SECOND EDITION.

The first edition of "The Rise, Fall and Revival of Dental Prosthesis" was, with the exception of one hundred copies, entirely exhausted in Chicago. This unexpected large sale has agreeably surprised the compiler, and believing that this immediate market was a criterion of its popularity, a second edition was the natural outgrowth.

Having noted carefully the imperfections that revealed themselves on a studious and painstaking review of the former edition, the necessity of a thorough revision of the whole body of the original text became apparent; and though this involved extended research and much labor, neither have been spared in the effort to render the present volume a faithful exponent of the evolution of dental art and science.

In the work of expurgation and amendment of the first volume I have availed myself of many of the suggestions made in various reviews of the original book, and take this opportunity and occasion to express my sincere thanks for the consideration shown toward it, both by the reviewers and readers. The revision it has undergone will, I hope, make it worthy of the continued patronage of those who have hitherto shown it such liberal encouragement. None knew better than myself how numerous were its imperfections. The manner in which they have been overlooked has served to convince me that those who were judges of the art and science, and could deal authoritatively, were disposed to encourage any attempt at its improvment, even though such attempt were marked by numerous shortcomings. Doubtless the kind reviewers saw that the book aimed at much more than was directly expressed upon its pages. At the best, therefore, such undertaking, of embracing the whole history of Dental Prosthesis, as viewed by a single individual, must needs be in some particulars unsatisfactory, if anything like a rigorous criticism be applied.

It may be truly said that the interest of the dental profession at the present time requires that the encouragement this work has received, should be extended to every undertaking of its kind. I hope that the success which has in this manner attended my labors may prove a stimulus to others who may devote themselves with possible better results to a similar task.

To those two reviewers, who, after having hastily

scanned the first edition, and then with anxiety sharply criticised its contents, I beg to say: that as professional reviewers, critics, you have done nobly, considering, you have never chanced to read, or presumably forgotten, "Pope's Essay on Criticism," in which able treatise are found these few words:

> " Whoever thinks a faultless piece to see
> Thinks what ne'er was, ne'er is, nor ne'er shall be."

I am fully conscious of existing defects in this work. The word "perfection," as we are aware, occurs in the " Book of Nature " only.

So far as facts are concerned, their particulars, and source, I have specifically indicated the proper credit and authority in their appropriate places throughout the body of the work. It would be most unjust to conclude this preface without publicly acknowledging the great obligations which I owe to scores of eminent men in the profession. Many of these beacon lights have contributed valuable suggestions, intelligent criticisms, and even several have supplied lengthy correspondence.

Trusting the present volume may be found to reflect with reasonable fidelity the present advanced state of this department of Dentistry, I respectfully submit it to the profession in the hope of its continued confidence.

Very Sincerely,   B. J. C.

CHICAGO, March 24, 1893.

.

SALUTATION.

"God bless those Surgeons and Dentists!
May their good deeds be returned upon
them a thousand fold. May they
have the felicity in the next
world to have successful
operations performed
upon them through
all eternity."

WASHINGTON IRVING.

INTRODUCTORY REMARKS.

An Historical Review of Dental Prosthesis.

Our happy fortune to live in an age whose masterpieces of accomplishment in art, science, industry and commerce, put to shame the extravagant fictions of oriental tales and the wonders ascribed to the gods and heroes of ancient mythology, must ever inspire us with grateful satisfaction. The changes produced by recent investigations and discoveries are so vast and appear so rapid that it is impossible to follow them and comprehend the power and thoroughness of the transformations that are daily taking place in the world around us. The application of steam and electricity astonish us by their wide range of influence on the conditions and relations of men; the ease and speed of movement and intercourse, constantly increasing, are ever putting us in new and unfamiliar situations. We have hardly accustomed our thoughts and habits to the one, before we are hurried on into the other. The faithful and abundant light shed by

science and the press does not suffice to keep our
minds fully informed of the rapid progress that goes
on in all departments of human life.

It is plain that we have entered on a new era,
the most extraordinary and momentous the world
has ever seen. The old and imperfect is being
uninterruptedly torn and cleared away and everything thoroughly reconstructed. The explanation
is, that we are now setting up the grand temple
of civilization, the separate stones and pillars of
which each nation and age has been commissioned
to hew and carve; and Father Time has requested
that all this grand masonry be left at the quarry
to await the time when, all the material being ready,
the master builder, America, should collect all the
scattered parts and raise the whole edifice and
designate the gorgeous structure "The World's
Columbian Exposition"—this to the astonishment
and joy of mankind.

All the institutions and structures of the past
may be considered temporary, erected in haste from
the material nearest at hand, not for permanence
but to serve the present time, while the special task
of the nations of this age is being performed. The
races nearest the birth of mankind worked on the
rougher parts of the ideal edifice that enter into the
stable foundations; those grand races, the Egyptians, Grecians, Etrurians and Romans, furnished
the noble outline which our modern humanity per-

fected by supplying what still lacked completion and adornment. America was reserved, designedly, for so many ages to furnish a suitable and unencumbered location for these central halls, mighty pillars and towering mirrors of perfection.

We begin to see that time, thought and experience have not wrought in vain; that progress is not phantom of the imagination ; that the human race is essentially a unit ; that civilization has been growing through all centuries, and is now approaching the prime of its manhood. The energies of all the peoples of the earth are prepared to exhibit unheard of achievements. The world was never so completely and so wisely busy as now, and America stands out in bold relief as the hospitable "Goddess of Progress." Within these halls of man is reflected the true status of human accomplishment.

In dentistry, the advances in the art and science during the past forty years have certainly surpassed the progress of most, if not all, of the older professions, shall here be most accurately portrayed and worthily represented. In this upward movement the members of the profession have borne their part. All honor to the skillful practitioners who have studied, and thought, and planned to uplift their profession and benefit their patients and themselves by devising improved methods and better instruments and appliances. But what they did, alone could not have caused the rapid pro-

gress of which dentistry boasts.   It needed the co-
operation of the merchant and the manufacturer to
put the improved methods and appliances where
they would benefit the profession at large: commercial judgment as to what was likely to be approved and demanded; commercial courage to
back up this judgment by investing money freely in
what that judgement approved; manufacturing
skill to know and enforce the highest attainable
standard of excellence in whatsoever passed
through its hands; and, last but not least, commercial methods to introduce successfully what
genius had invented, commercial judgment approved, and manufacturing skill produced.

Of the grand strides of progress few branches
of science and art deserve a higher rating and
garner a more lasting glory than Dental Prosthesis,
"the mother of longevity."

To the faithful dental student, who is ever
yearning for fresh draughts of information, every
subject that pertains to the history and progress of
dental art and science is fraught with the deepest
interest, and probably no feature in the annals of
dentistry solicit his attention sooner, and merit
such sincere consideration as does the story of the
rise, fall and revival of the dental art.   The dentist
who is ignorant of the beginning of the trials and
tribulations of the early surgeon-dentist is not
unlike the patriot who glories in the triumphs and

achievements of the native land, but knows not of
the making of the established institution he so
fervently loves; such patriotism is but local and su-
perficial at that.

There is an Arabian maxim of much truth which
says: "If you are about to acquaint yourself with
a man, first learn where he was born, and next
how he was raised." And this good advice in
sentiment might well be given to those who are
about to acquire a knowledge of the mysteries of
the arts and sciences, namely: first learn of the
origin of the art or science and next determine its
progress.

The dental practitioner, on the contrary, usually
toward the end of his professional career, learns of
the latent beams of dental history while, had he
gleaned the information in his initiatory studies
of the art and science, he would have more fully
appreciated the modern status and advancement
of his calling. But the old saw, "Better late
than never," is quite applicable, and hence in this
small tome the author solicits the attention of both
the already learned, and the ambitious beginner,
and trusts these fragmentary parts of the dental
history, as presented here in an unbroken chain,
may call forth a high admiration for, and sincere devotion to, Dental Prosthesis.

In order that we may more clearly understand
the historic narrative, let us see if we all agree as

to the meaning of the term Dental Prosthesis.   No
doubt we are all harmonious when we analyze the
first word Dental, and say it is of Latin origin, and,
as applied here, is the simple adjective form of the
Latin noun "dens," a tooth.   But we are apt to
differ materially as to the rendering of Prosthesis.
Present time dictionaries tell us that this word is
of Greek derivation and signifies, add to, replace,
affix, or restore.*   Thus the term Dental Prosthesis
is a combination of two words, the one of Latin,
the other of Greek origin, and when connected in
their literal sense designates "tooth addition,"
"tooth replacement," "tooth affixion" or "tooth
restoration."   Dental Prosthesis would seem to
imply nearly all dental operation, since little else is
the dentist called upon to do other than replace,
add to, affix and restore dental organs.   This
would mean that all specialities of dentistry are
branches of Dental Prosthesis,† which, in fact, is
the case, though not generally so accepted.
Hence if the dentist fills a crown cavity with foil
or plastics he simply restores the dental organ by

*These dictionaries: Century, Encyclopaedic, Webster's, Inter-
national, Thomas' Medical, Dunglison's, Zell's Encyclopaedic and
Greek-English.
†Dental Review, vol. V., p. 438.
Proceedings of Illinois Dental Society, 1891.
Address—Magill before Illinois Dental Society, 1891.
Dental Cosmos, New Series, vol. XI., p. 315.
Ibid, vol. XXVI., pp. 180, 181.
Ibid, vol. XXII., p. 1004.
Proceedings of American Dental Association, 1884—"Distinctive
Names and Phrases."

replacement ; if he crowns a root with porcelain or
gold-shell crown he again resorts to prosthetic
art, and restores by addition ; if he replant or
transplant a tooth or teeth, he further follows prosthesis and restores by replacement ; if he attach
to the natural teeth several artificial ones by means
of a system of bridgework, he but restores the denture by addition ; if he treat with the aid of medicinal agencies a sore tooth, he similarly labors in a
prosthetic sense, since he restores the tooth to its
natural health; * if he is sought to relock a jaw or
aid in healing a fractured maxilla, he replaces and
restores, thus again adopting prosthetic principles.
There are but few operations of which I now think,
that a dentist's services are sought, which are not
indirectly prosthetic, and among these are the devi-
talization of a nerve, the administration of an
anaesthetic and the extraction of a tooth.

Latest authorities pronounce the custom of saying Prosthetic Dentistry as incorrect, and should be
designated Dental Prosthesis; "tooth replacement,"
not "replacement tooth," as in the former remark. †

The antonym of Dental Prosthesis is Dental
Aphaerisis, which implies, to "take from," "omit,"
"remove" or "subtract," i. e., the extraction of a

---

*Prosthetic Hygiene—Prof. J. Hall Lewis, pamphlet, 1890.
    Journal Fuer Zahnheilkunde. vol. VII., p. 180—"Prosthesen
Heilung."
    †We shall see the rational reason for this when we study, later in
this work, the terms Mechanical Dentistry, viz: Dental Prosthesis.

tooth or removal of an ulcer or tumor. Hence a fair definition for prosthetic dental art would be as given by Dr. Harris :

"Dental Prosthesis is that branch of dental science which teaches the art of replacing lost organs of the mouth or any part thereof ; it includes the laws and principles which determine and regulate the processes employed in the construction of all forms of dental mechanism ; also the properties and relations of all materials used in these processes. Replacement and therapeutics are its distinctive dental peculiarities." *

Correctly spelled, the word has the following letters—p-r-o-s-t-h-e-s-i-s ; † but of late it seems permissible to render it p-r-o-t-h-e-s-i-s; this latter custom, however, is not in harmony with the true acceptance of its specific use relative to dental art. The Greek preposition "pro" signifies "before" or "forward," and the root word "thesis" in Greek means "to place," hence, according to rules of synthesis, we would determine the word to mean "to place before;" for example, in the science of philology we use the term "prothesis" with such intent, when in synthesising we prefix or "place before" the root word a syllable as be-loved and re-turn. ‡ On the other hand the word "prosthesis"

*Harris' Principles and Practice of Dentistry, 1889, p. 715.
†Unabridged dictionaries of present time.
Medical and Dental catalogues of the Columbian World's Fair.
‡Dictionaries: Century, Encyclopaedic and Zell's.

serves us better, since we find that the prefix "pros" in Greek signifies "to add" or "to put to;" and hence joined to the substantive, "thesis" (to place), signifies "to put in place" or "replace," which latter form fully answers the latent meaning.

When the word is spelled p-r-o-s-t-h-e-s-i-s, it has three syllables, pros-the-sis, and may be pronounced either with primary accent on the first, and secondary accent on the second syllable; or primary accent on the second, and secondary on the first syllable; in both instances all vowels except "e" have the short sound; the latter pronunciation, by way of euphony, is the favorite.

Dental Prosthesis, as an art,* has been practiced for ages. Dentistry is generally considered a modern science, but on careful investigation we find, on the contrary, that it is ancient, and there is abundant evidence to show that the art is of great antiquity. Although it is less than a century that it has taken the rank of a distinct profession, attention was directed from the earliest period of civilization to the means of preserving and improving the beauty of natural teeth. In order that all possible doubt be eliminated, as to its antiquity, we will give it careful consideration, and in no small measure profit by the research.

Fortunately, we are not confined to mere tradi-

---

*Why Dental Prosthesis is considered both an Art and a Science will be shown later in this work.

tion nor to ocean-lore for evidences of vanished races, and the wonderful monuments of their accomplishments, or the brilliant stages of civilization. After carefully perusing the story of the ancients we learn that much credit belongs them as inventors, promoters, students and masters; and that much which we claim to be purely modern conception, on close study proves to be ancient, most ancient.

If we seek for historical knowledge of any important invention, discovery or attainment, we must conduct our investigations into the remote past, in which are buried marvelous secrets that ages ago perished with their possessors. A thousand illustrations might be introduced in proof of the claim that civilizations rise, fall and revive like the tides of the sea; for "human progress," says J. W. Buel, "is so intermittent that its mutations are like the motion of a pendulum, swinging now across the valley of benighted barbarism and up the gentle slope toward the pinnacle of exaltation, then driven back by adverse influences, scourges, devastating wars, immortalities, until, gaining momentum, it crosses the shadowy abysses and rises to the peak of human discouragement. Here the pendulum pauses until the gravity of ambition pushes it again forward, thus imparting a reciprocal impulse which keeps it in perpetual motion."

How applicable this quotation is to the progress of dental art and science, the following pages we hope may tell.

Of the origin of the art of dentistry no one can speak with certainty, as its early history is shrouded in the mists of oblivion, but dental operations we learn, and on most eminent authority, are recorded in very remote times.   It is impossible to determine the native home of Dental Prosthesis, but in all probabilities Egypt, the most highly civilized nation of the ancient world, claims the art as a cherished creation.

EGYPTIAN DENTAL ART.

Before entering into a detailed account of Egyptian dental art it is necessary to remind the student of ancient history of a few latent facts, in order that liberal allowance may be given to ancient traditions, narratives and history.

In the year 332 B. C. Alexander the Great founded in Upper Egypt a city which he appropriately named Alexandria, and made it the capitol of Egypt. Alexander had been a faithful student under Aristotle, and it was his ambition to establish in his newly planned city the largest library and museum of the world. He, however, died a few years subsequent to the founding, had he lived a score years longer his anticipations would have been realized, for we learn that in 321 B. C. the library and museum contained upward of 700,000 volumes, and as many more rolls of papyri, thus the most remarkable collection of the ancient world,* containing the literature, art and science of Greece, Italy, Phoenicia, Arabia, India, China and Continental Europe. In connection with this voluminous library was the museum in which, at public expense, all the students of the world gathered and studied under the imme-

---

*Encyclopaedia Britannica, vol. I., p. 498.
American Cyclopaedia, vol. I., p. 291.
Peoples' Encyclopaedia, vol. I., p. 58.

diate instruction of the most eminent scholars of
literature, theology, art and science.* This museum
or academy of science was in many respects not
unlike a modern university. The Alexandria
library at that time was, in truth, the depository of
all the written thoughts of man. Money was
lavishly spent in order that the library be amply
provided with means for acquiring information, and
under the watchful eye of "monarchs of learning"
the world was enlightened and the progress of
civilization was marked by the term, Alexandrian
Age. The names of Euclid, Hipparchus, Clement,
Origen, Theon, and his daughter Hypatia, and
many others of equal distinction, shed their glory
upon the literary reputation of grand Alexandria.
With such advantages as these is it any wonder
that this institution gained a world-wide renown
and exerted such an ennobling influence on mankind?

But what interests us as dentists more in particular is, that Alexandria was especially distinguished for her medical, dental and optical schools.†

---

*Zell's Encyclopaedia, vol. I., p. 60.
Encyclopaedic Dictionary, vol. I., p. 131.
Century Dictionary, vol. I., 137.
Alexandria and Her Schools—Kingsley, 1854.
History of Alexandrian Schools—Matter, 2 vols., 1844.
Alexandrian Schools—Simon, 2 vols., 1845.
†Herodotus, vol. II., p. 84.
Peoples' Library of Information—Washburn, 1876, p. 269.
Dental Cosmos, vol. X., pp. 346, 347, 348.
Items of Interest, vol. XII., p. 253.
L. C. Ingersoll, Methods of Dental Education, 1890.

Here lived and labored Herophilus, Galen, Aetius, and many others who adorn the early annals of medical science. The splendor and glory of this hospitable seat of learning did not remain unshaken, for in 80 B. C., Ptolemy Alexander, a weak and vicious monarch, bequeathed the city and its valuable collections to the ambitious Romans, under whom the city as an enlightened educational centre rapidly declined. Notwithstanding the removal of many of the most precious works of art and science to Rome, its greatness continued to grow until 30 B. C., when Julius Caesar waged unrelenting war, and the library and museums were partially destroyed by fire. The final work of destruction took place in A. D. 642 by the Saracens. Amrou, the commander of the army of Omar, was disposed to spare the library, and wrote to the Caliph to obtain his consent, but the bigoted Mohammedan wrote back his well known reply: "If these writings of the Greeks agree with the Koran, or book of God, they are useless and need not be preserved; if they disagree they are pernicious and ought to be destroyed." The sentence of destruction was executed, and accordingly, it is said, the books were employed to heat the 4,000 baths of the city; such was their number that six months were barely sufficient time for the consumption of the precious fuel.*

---

*The Histories of Alexandria, as quoted in preceding pages.

Thus the intended legacy was consigned to the
angry flames, and the loss is inestimable. This
conflagration nearly completely burned the authentic history of antiquity and left tradition prime
factor on the field of learning. In consequence of
this destruction we moderns are much at the mercy
of narration and similar weak authorities; but,
happy still, the entire records of man were not
destroyed, since when the Romans acquired the
Alexandrian library they carried multifarious volumes, records and works of art and science to Rome.*
Through these latter archives, and in addition the
minor records found throughout the ancient king-
doms and empires, we manage to gather much that
proves authentic. From what has been preserved
we learn that the Egyptians cultivated the science
and art of medicine at an early date, each physician applying himself to some one specialty.†

"Man ever since his creation has been subjected

*American Cyclopaedia, vol. I., pp. 290, 291.
  Peoples' Encyclopaedia, vol. I., p. 85.

†Herodotus—Euterpe, p. 84.
  Travels and Correspondence of Dr. C. A. Kingsbury—Pamphlet,
1863.
  Odontographic Journal, vol. I., pp. 86, 87.
  Dental Cosmos, vol. X., p. 348.
  Items of Interest, vol. XII., p. 253.
  Dental and Oral Science in America—Dexter, 1876, p. 1.
  Archives of Dentistry, vol. II., p. 88.
  Transactions of Odontological Society of Great Britain, vol. VII.,
N. S., p. 239.
  History of Alexandrian Schools, quoted on preceding pages.
  Peoples' Library of Information—Washburn, 1876, p. 269.
  Dental Review, vol. III., p. 435.

to disease and that necessity, which has ever been the mother of invention and discovery, must have early taught him to use some means for its alleviation or cure, rough and uncouth perhaps, still in a measure answering his purpose. Thus we find ancient writers referring to the practice of dentistry as being coeval with the birth of medicine. Modern research has conclusively demonstrated the fact that dental caries was prevalent in those ancient days, and that this disease also received due and marked attention at the hands of the early specialists of medicine. The Egyptians, as well as their border neighbors, divided the pains and ills that affected different organs and members of the body into different classes, not knowing that many disorders originated in the same locality. Thus they began to study human ailments; each practitioner devoted his time to one class of disease having its existence in one portion of the body. Thus there sprung up oculists, aurists and dentists. But in as much as the teeth were not so subject to affection as other organs the dentists were neither so plentiful in those days nor so elevated in repute." *

Thus away back in dim centuries, when mythology had its happy reign and historians began to chronicle in their order transpiring events, we find dentistry was studied and practiced with great success. There can be little doubt that sunny

*New York Dental Journal, vol. I., p. 5.

Egypt was the birthplace of ancient dentistry; at any rate the Greek historian, Herodotus, cites that the Egyptians practiced dental art. In his second book narrating his travels through Egypt he states that the art and practice of medicine was divided among the Egyptian priesthood, each physician or surgeon "applying himself to one class of disease only; some to the head, others to the eye others to the teeth and even others to internal disorders."

And although little is known of the attainments of these ancient practitioners of dentistry, judging by the work deposited in some of the tombs of Egypt, one would willingly conclude that the practitioners of that time were comparatively learned and fairly proficient in the science of Dental Prosthesis.*

The great Egyptologist, Ebers, has proven that in the Egyptian medical schools they had special teachers of dentistry upwards of three thousand years ago.†

Had Egyptian universities their chairs of dental surgery? If so, an added laurel may deck the fame of the "land of science and sacred recollections." The following from the eminent and reli-

---

*Surgery Paulus Aegineta (translated by Francis Adams for the Sydenham Society), 3 vols., 1844.
    Dental Art—Dr. G. Carabelli, 1 vol., 1844.
    Dental Review, vol. III., p. 435. Article by Dr. John J. R. Patrick.
    Archives of Dentistry, vol. II., p. 83.
    †Paulus Aegineta, vol. VI., p. 10.

able historian, Read,* goes far to establish these facts:

"The physicians were compelled to prescribe for their patients according to rules set down in the standard works. If they adopted a treatment of their own, and the patient did not recover, they were put to death. Thus even in desperate cases heroic remedies could not be tried, and experiment, the first condition of discovery, was disallowed.

"It is one of the first axioms of medical science that no one is competent to treat the diseases of a single organ unless he is competent to treat the diseases of the whole frame. The folly of dividing diseases of such organs as the head and stomach, between which the most intimate sympathy exists, is evident to the unlearned. But the whole structure is united with delicate white threads and innumerable pipes of blood. It is scarcely possible for any complaint to influence one part alone. The Egyptian, however, was marked off, like a chess-board, into little squares, and whenever the pain made a move a fresh doctor had to be called in.

"Even their books (now in the library of Thebes), so few in number, were not open to all the members of the learned class. They were the

---

*Martyrdom of Man—Read.

manuals of the various departments or professions, and each profession stood apart; each profession was even subdivided within itself. In medicine and surgery there were no general practitioners. There were oculists, aurists, dentists, doctors of the head, doctors of the stomach, etc., and each was forbidden to invade the territory of his colleagues. This specialist arrangement has been highly praised, but it has nothing in common with that which has arisen in modern times."

The various specialists of medicine and surgery, according to Rev. William Smith, were paid their salaries from the public treasury, and thus indirectly the rich paid for the medical service rendered to the poor.

Another writer * has the following interesting catalogue of evidences: "Some (physicians) made the treatment of the dental organs their special branch of study; and although we are unable from the records that have come down to us to obtain a clear and satisfactory knowledge of the exact condition of dental science and art at that early period, we have no difficulty in tracing our profession back to the days of the Egyptians, through the medium of historical records, as well as from the existing evidences and specimens of dental art discovered now and then in the mummified bodies taken from the tombs and catacombs of Egypt.

---

*Travels and correspondence of Dr. C. A. Kingsbury—pamphlet, 1868.

I have met several gentlemen, whose veracity I could not question, who stated that they had not only seen artificial teeth, but even gold fillings in the teeth found in the sarcophagi of the ancient Egyptians. It will be remembered that the Egyptians attached great value to the dental organs and one of their most severe punishments consisted in having one of the front teeth extracted. It would be natural to suppose that in order to avoid suspicion of guilt, as well to restore the loss, artificial teeth were invented and substituted for the lost organs."# •

Exhumed, from the time worn Egyptian tombs antedating the records of Herodotus, mouldering skeletons present arrays of gold filled teeth;† and able authority, states that the art of clasp-work was understood to some exactness.

---

*Dental Cosmos, vol. X., p. 348.

†Dental Cosmos, vol. VIII., p. 607.

Dr. C. A. Kingsbury's Correspondence—1868.

Western Dental Journal, vol. I., p. 294

Archives of Dentistry, vol II., p. 83.

American Cyclopaedia, vol. VI., p. 21.

Peoples' Library of Information, 1876, p. 270.

Dental and Oral Science in America, by Dr. Dexter, 1876. pp. 1 and 2.

British Journal of Dental Science, vol. XXXI., p. 77.

Ibid, vol. XXXI., p. 88.

Items of Interest, vol. XV., p. 241.

Research of Dr. Mellersh.

American Journal of Dental Science, vol. IX., N. S., p. 45.

Dentistry Among the Ancients—Chap. A. Harris, 1889.

Johnstons' Dental Miscellany, vol. VIII., p. 80.

British Odontological Society—Ancient Dentistry, 1881.

Popular Account of the Egyptians—Sir Gardener Wilkinson, D. C. S., F. R. S. S., vol. II., p. 350.

Crania Egyptiaca—Morton, 1844, pp. 10, 25.

† Items of Interest, vol, XV., p. 228.

In many of our present day medical and dental
journals we are informed that in some of the royal
mummies * taken from the catacombs of Egypt, sets
of artificial teeth were discovered † in which the
plates were of wood carved to fit the roof of the
mouths, while the teeth, which were of brass, were
ingeniously attached.   Charles Creighton, the
eminent authority on the history of general surgery,
says this of the ancient Egyptians, relative to their
dental and surgical skill: "Cupping vessels made
of cow-horn have been found in ancient Egyptian
tombs.   On monuments and walls of temples are
figures of patients bandaged and undergoing operations at the hands of the surgeons.   In museum
collections of Egyptian antiquities there are lancets,
forceps, knives, probes, scissors, surgical instruments for the ear are figured, and artificial teeth
have been found in the mummies."

Gold work was understood by these ancient
practitioners. A set of artificial teeth was discovered, the base of solid gold and the teeth of
ivory. ‡   One of the mummified bodies of an Egyp-

*"It has been estimated that more than 400,000 human mummies
were made in ancient Egypt. Sepulchres have been opened in which
thousands of them were found deposited in rows, one on another, without
coffins. Shiploads have been transported to England, and ground up for
use in fertilizing the soil." John J. Anderson, historian.

†Historical Researches upon the Dentist's Art Among the Ancients
—Dr. J. R. Duval, Paris, 1808.

‡Narrative of the Operations and Recent Discoveries Within the
Pyramids, Temples, Tombs and Excavations in Egypt and Nubia--
Battista Belzoni, 2 vols., London, 1822.

tian Pharaoh demonstrates most conclusively that
natural teeth were not only well cared for in the
way of gold and lead fillings, but that the aged
Pharaoh had, perhaps by accident, lost one of his
incisors, and the court dentist did the best he
could and carved an ivory tooth of similar shape to
the lost organ, and by means of silk ligatures
fastened it to the adjoining sound ones.* Bone
and wooden teeth were frequently discovered by
Belzoni in his researches in Egypt some years
since.

A writer in one of our early dental journals†
adds: "The instruments used, appear to have been
clumsy and illy fitted to the delicate labor they had
to perform. Among the strange frescoes and base
reliefs found at Thebes and also at Memphis, are
many representing the daily labors, the trades and
the professions of the ancient Egyptians ; and one
of these portrays a dentist operating on a patient.
Although the knowledge of dental science cannot
be considered to have greatly advanced among the
Egyptians, still there are evidences existing which
justify us in giving them credit for a greater degree
of skill than we suppose could have existed at that
early period. * * * Some of the mummies
which have been exhumed from the ruins of ancient
Egypt have presented palpable evidence of having

* Papyrus **Ebers** (1874). Egyptian Medical Art—G. M. Ebers.

†New York Dental Journal, vol. I., p. 6.

had their teeth filled with gold and various other substances. We do not, however, think that implicit reliance should be placed upon the statement that, because these bodies were found with gold and aromatic preparations in the teeth, it follows that filling teeth to preserve them was customary in life.

"The object of the Egpytians in embalming bodies was evidently to arrest decay and, as far as possible, to preserve the existing appearance of the corpse. With this end in view they (the Egpytians) would naturally use particular care with such members of the body as had already commenced to decay.* This appears to us a feasible and probable view of the matter, and we lean toward it rather, that none of the other medicinal operations of the Egyptians are of a nature to induce us to believe that they were far enough advanced in science to undertake the filling of the teeth during the vitality of the body. This belief does not, however, prevent us from giving them their due credit, for knowledge of the chemical properties of the substances they used, a knowledge which in perfection at least is now reckoned among the lost arts."†

Whether these gold fillings were inserted prior

---

*The writer, we perceive, was of the belief that the embalmer and not the dentists filled the teeth with gold. This view may be rational but we are at a loss as to a strong critique.

†Dental Science in Earliest Ages—New York Dental Journal, vol I., p. 4.

Ibid, vol. II., pp 2 and 3.

or subsequent to the person's death seems beyond the judgement of present humanity. It is evident that some knowledge of gold foil was extant and that Dental Prosthesis likewise was a familiar art. A writer in a recent dental journal* says: "While it is true that unmistakable traces exist in the literature and discoveries in Egypt of some valuable observations on the dental system, and the substitution of artificial teeth for the absent natural ones, it must be borne in mind that those notions and productions must of necessity have been largely of an empirical and rudimentary character. Indeed, it were a curious contradiction to our established ideas of a more or less systematic evolution of the different sciences to assume otherwise."

The museum in Liverpool, England, contains, besides artifical teeth, "two teeth of sycamore wood set in gold." Home museums and collections have many specimens of "the mode of fastening by ligatures or bands of gold or silver wire tying the substitute to its neighbor."

It appears that women, too, practiced the art of medicine and dentistry in those ancient days, since we find that "Women, being forbidden to consult with men, received services from their own sex."† At the present time there are scores of women dentists practicing dentistry in Egypt.

---

*Items of Interest, vol. XV., p. 241.
†Correspondence of Dr. Edward Warren—1874.
Scientific American—1874.

Egyptologists are still at work deciphering the
ancient written rolls of papyri, and much information remains yet to be derived; the explorations
are disclosing many interesting archaeological remains and treasures ; the large museum of Egpytian
antiquities at Boolak, Egypt, is rapidly being filled
with the treasure-trove, and before another decade
the world will be, through the medium of the Egyptologist, more definitely informed of the advanced
civilization of the "Mother of Arts and Sciences." *

ADDITIONAL BIBLIOGRAPHY.

Manners and Customs of the Ancient Egyptians
—Wilkinson, 5 vols., 1847; History of Egypt from
the Earliest Times to the Conquest by the Arabs—
Samuel Birsch, 1846; Medical Papyri of Ancient
Egypt—Birsch, 1871; Ueber die Medizinischen
Kenntnisse der Alten Aegypter—Brugsch, 1853;
Aegyptens Stelle in der Weligeschichte—Bunson, 5
vols., 1845; Ancient Egypt under the Pharaohs—
Kenrick, 2 vols., 1850; Handbuch der Gesammte
Aegyptischen Alterthumskunde—Uhlemann, 4 vols,
1857; Rawlinson's Herodotus, 1858; Euterpe, by
Herodotus; Zur Aegyptischen Forschung Herod-
ots—Buedinger, 1873; Museum of Antiquity—
Yaggy, 1885.

---

*See Appendix for additional notes.

HEBREW DENTAL ART.

This people hardly deserve a separate chapter, as most of their knowledge of the teeth was borrowed, but since they left testimony of their skill it may be justifiable to record. The ancient Hebrews were not noted for having, at any period of their existence, displayed any great amount of mechanical ingenuity and originality in the arts and sciences and can not with good authority be accredited. These people if at all gifted in this direction they have failed to leave a lasting impression. Language and law were perhaps their great triumphs. In order to learn much of the inner or private life of the Hebrews one must be a Talmudist, for in the great Talmud the requisite daily life and habitation can be accurately traced; in this standard record many laws concerning the natural and artificial teeth can be found.

The ancient Hebrews too well knew the worth of teeth, and the great King Solomon wisely nicknamed them the "millstones," and were by his people recognized as the motive power of life. Moses legislated his famous law known as "tooth for tooth,"* an edict which was altered and explained in the Talmud to mean a fine or penalty.

---

*Dentem pro dente—Exode, ch. XXIII.

The man who broke the tooth of his fellow-man
had to pay the unfortunate a sufficient amount of
money* for damage done, or have the substitute
placed at the cost of the aggressor.†

So highly did the Hebrews value a natural
tooth that if a master broke the tooth of a slave
the latter was set at liberty on account of his great
misfortune.   Mosaic law in this respect was in
force long after the Christian era, as the following
from Talmud says:‡ "Rabon Gamliel, who was
the teacher of the Apostle Paul, had a slave, Tabi
by name, and as he was anxious to set him at
liberty forever he broke his slave's tooth that the
latter should be free, and the Rabbi was so happy
the day of his slave's liberation that he gave a
banquet to his friends, besides sending his slave
off with a present."

The Talmudical Folk-lore says: "If a man
dreams that his false teeth have fallen out it is a bad
omen that his children will soon die."   Indeed the
dental profession was in a state of semi-cultivation
under the care of the learned sages of the Talmud,
and modern dentists would be astonished to learn
that the art of replacing false teeth for natural

---

*"That a fine of twelve crowns should be levied against any person
who had broken the tooth of his neighbor."—Jewish Law, fifteenth cen-
tury.
†The Youth's Dentist—Dr. J. R. Duval, Baltimore, 1848, p. 60.
‡Items of Interest, vol. XIV., pp. 415, 416.
 Talmud Babli.
 Dentistry Among the Hebrews—Naphthali Herz Imber, 1892.

ones was practiced among the Hebrews more than two thousand years ago.

Samuel, the dentist who lived after the destruction, was the house physician and dentist of the famous Rabbi Jehuda, the Saint. The latter was often afflicted with toothache but was cured by Samuel. What drug this Rabbical dentist used is unknown, but according to the Talmudical narration it must have been chloroform or something with similar qualities, since the Rabbi's suffering was stopped by the use of inhalation of anaesthetic fumes.

Among the Hebrews it was strictly forbidden by law to carry jewelry or precious metals or stones on the Sabbath, but the Talmud wisely allowed the Jewish women "to go out on the Sabbath with her false golden or silver teeth." Some Rabbis allowed their people to wear the silver false teeth since these appeared natural, but the use of golden false teeth on Sabbath were prohibited.*

An authority on Hebrew customs and ceremonies says:†

"Among the orthodox Jews, especially of the large cities of Europe, where the Rabbis are regarded as the law givers of Hebrew communities, even now after the Ghetto era, none will submit to a dental operation unless the ingredients used by

*Rabbianical Code.
†Rev. H. J. Sharp.

the operators are pronounced by their spiritual advisers as 'not prohibited' by the Rabbinical code or the ceremonial law. Anything obtained from the bodies of such animals as swine, hippotami, oysters, etc., would be positively forbidden to be used in dentures to be applied to Jewish patrons."

In later times the artificial teeth were made of ivory instead of wood, and this statement is substantiated by the curious fact that the Hebrew term tooth is "shen," while the term for an elephant is "shenhob." Now it is highly probable that the term for tooth was derived from the word elephant, as they used the latter's white tusks in making false human dentures.

Hebrew dental art is so intimately interwoven with the Egpytian that aside from the Talmudical obligations, the dental art was practiced and understood by Hebrews and Egyptians alike.

ADDITIONAL BIBLIOGRAPHY.

Talmudical Commentary—Don. I. Abarbanel; Research In Antiquity—R. Waller; Critique on the Talmud—Perl N. Krochmal; History of the Hebrews—Zunz; Talmudicum et Rabbinicum—Buxton; Hebrew Antiquities—Sully; Geschichte der

Israels—Wellhausen ; History of the Hebrew Monarchy—Newman ; Encyclopaedia der Classischen
Alterthumswissenschaft—Pauly ; Egyptian and Hebrew Antiquities—Young, Bunsen, Letronne and
Champollion ; Transaction of Society of Antiquities
of France ; Hebrew Arts and Sciences—Dean Stan-
ey.*

CHINESE DENTAL ART.

The Chinese, it must be remembered, were in ancient days a persevering people and made wonderful advancements in the arts and especially in the sciences. In medicine and surgery they were considered fairly advanced. One of the ancient kings of China instituted a free medical school for those citizens who were inclined to study medicine or its many branches.* Although it is very difficult to obtain facts and figures of ancient China it, notwithstanding, fortunately happens that a Chinese MSS., deposited in the French Academy of Science, gives us much information as to Chinese methods of dental practice. In this paper we are told, among other things, "that the court dentists ever preserved for the royalty the entire natural denture and keep the same 'clean and sweet.'" The paper further states "that the dentist used a peculiar pitch of a white appearance, and this was used to restore decayed teeth."

Of animal physiology† and medicine the Chinese have very crude notions, as is shown by their various ideas of the human body, and their practice

---

*Encyclopaedia Britannica, vol. I., 789.
†Peoples' Encyclopaedia, vol. I., p. 427.
American Cyclopaedia, vol. XI., p. 345.

of medicine and surgery must of necessity be empirical.*

The practice of dentistry in China is doubtless very ancient, but it has not attained that perfection which characterizes the modern art.

It is well known that the Chinese attribute toothache to the gnawing of worms, and that their dentists claim they take these worms from decayed teeth. How this is accomplished is as follows: "When a patient with toothache applies for relief, if the tooth is solid in the socket, the gum is separated from the tooth with sharp instruments and made to bleed. During this operation the cheek is held to one side by a bamboo spatula, both ends of which are alike, and on the end in the hand some worms are concealed under thin paper pasted to the spatula, the paper being the same color as the spatula. When all is prepared this end is adroitly turned and put into the mouth, and the paper, becoming moistened, is easily torn with the sharp instrument used for cutting the gums, gives up its worms, which mix with the bloody saliva, and the dentist leisurely picks them out with a pair of forceps. The patient having ocular demonstration that the cause of his disease has been removed, has good reason to expect relief, which, in most cases, would be the result from the bleeding of the gum and the fright of the patient.

---

*Peoples' Encyclopaedia, vol. I., p. 427.

"When the toothache returns, as it will in almost every case—perhaps within an hour, or maybe not for one or two days—the patient again seeks his dentist for relief, and the same operation is performed, finding more worms, which of course explains the recurring trouble, and this is continued from time to time until the tooth ceases aching altogether of its own accord from the natural death of the pulp. Another fact might be mentioned : the standard medical books of China teach, and have taught for ages, the idea of worms in a tooth causing it to ache. The above practice is resorted to when the tooth is firmly set in the jaw, or is not so loose as to be removed with the fingers or by a slight force or pressure with iron instruments.

"The insertion of artificial teeth has been practiced in China for ages before it was introduced into Europe, and has one great recommendation, viz., cheapness. The material used is bone or ivory, and the tooth, having been sawed and filed into the proper shape, is fastened to the adjoining teeth by a copper wire or catgut string. If two or more teeth are required, they are made in one piece, and a hole drilled the whole length, through which a double string or wire is passed, which loops over the natural tooth at one end and is tied to the tooth at the other. This work, although rude in the extreme, is, as to looks, better than the absence of the teeth, and is of some use in mastication.

The cost of a single tooth will be from five to ten cents, and the charge for half a dozen would be from thirty cents to half a dollar. At these low rates all classes may avail themselves of the benefit, and those who practice the art do a thriving business." *

"The Chinese themselves do not believe in dissection, and there is no body-snatching here. They believe that the heart is the seat of thought, that the soul exists in the liver, and that the gall bladder is the seat of courage. For this reason the gall bladders of tigers are eaten by soldiers to inspire them with courage. The Chinese doctor or dentist ranks no higher than the ordinary skilled workman. He gets from fifteen to twenty cents a visit, and he often takes patients on condition that he will cure them within a certain time, or no pay. He never sees his female patients except behind a screen, and he does not pay a second visit unless he is invited. His pay is called "golden thanks," and the orthodox way of sending it to him is to wrap it in red paper. The dentists look upon pulled teeth as trophies, † and they go about with necklaces of decayed teeth about their necks, or with them strung upon strings and tied to sticks. Toothache is supposed to come from a worm in the tooth, and

---

* Report on Chinese Dentistry, 1877—Drs. J. G. Kerr and G. O. Rogers.
Dental Cosmos, vol. XIX., p. 382.
China Review, February, 1877.
†Dental Review, vol. V., p. 360.

there is a set of female doctors who make a business
of extracting these worms." *

Artificial teeth among the Chinese of medieval
times were seldom worn, since the dental surgeon
not only seemed skilled enough to preserve them,
but the Chinese were known to be the possessors
of sound teeth.† We now leave the superstitious
Chinaman and wander to his neighbor, the skillful
and dexterous Japanese.

ADDITIONAL BIBLIOGRAPHY.

History of Chinese Medical Art—Creighton ;
History of Medicine Among the Asiatic—Wise ;
Transactions of the Royal Asiatic Society ; Resume
of the Chinese Culture—Julien ; Medical and Surgical Advancement Among the Chinese—Medhurst.‡

---

*Travels of Dr. Frank Carpenter.
  Dental Advertiser, vol. XX., p. 131.
†St. Louis Dental Quarterly, vol. 1, pp. 13, 17.
‡See Appendix for additional notes.

JAPANESE DENTAL ART.

The following interesting account from one [*] who lives among the Japanese well illustrates the advancement made in dentistry :

"It is a little remarkable that a nation which places the value they do upon their teeth, and who take the care that is everywhere evident of their appearance should be ignorant of everything relating to them, other than their mere mechanical substitution.

"Taken as a race, the Japanese have not good teeth, neither can they be said to be very bad. Caries, with its resulting odontalgia, is quite common ; but the most frequent cause of trouble is the accumulation of tartar. To such an extent is this true that it constitutes the extracting agent of the aged. It is, indeed, rare to find an elderly person with teeth. As might be expected, the principal accumulation is behind the lower orals ; these are frequently entirely cemented together by a dense, dark brown deposit, of a quarter of an inch in thickness. But far more dangerous to the integrity of the organs is the gradual deposit around the necks of the teeth far up under the gums.

*W. St. George Elliott, M. D., D. D. S.,Yokohama, Japan.

"It is one of the peculiarities of this peculiar land that among the females one seldom meets with beauty, either among the very young or the aged; in both cases they are entirely devoid of color, but in early womanhood it is not rare, nor is color wanting to lend its charm. The teeth of the musmies, or daughters of Japan, are objects of envy, but the horrible custom of blacking the teeth after marriage destroys what little beauty time had not yet stolen.

"Irregularities are common. Their teeth being large, the jaw is not sufficiently expanded for their proper placement. Considerable care is taken to have the teeth appear white, the native brush consisting of tough wood, pounded at one end to loosen the fibres, when it resembles a paint brush; it is used with several kinds of powder, red, ash-color and white; they are all of a coarse structure, but answer the purpose very well with the soft, wooden brush employed. Owing to its shape, it is impossible to get the brush behind the teeth.

"In mechanical dentistry they rank far above any barbarian or semi-civilized nation. In point of fact, I believe they stand alone. Their denture answers admirably the principal object of one, i. e., the mastication of food. This, however, applies only to full sets. All full upper cases are retained in place by the atmospheric pressure. This principle is coeval with the art. The oldest inhabitant does not know when the principle was introduced.

"Owing to the fact that dentistry existed only as a mechanical trade, the status of those who practice it is not high. In a country where class distinctions are so marked, and where the laws prescribe the dress and social position of all, it is graded with carpenters, which fact is shown by their word, 'hadyikfsan' (tooth carpenter). Dentistry, to some extent, is practiced as an itinerant business. The carver, taking his seat beside the highway, exhibits his gilded sign, specimens and material. When not engaged in the all important business of gossiping he plies his chisel, shapes a denture, or grinds on a slab, a bit of quartz for a tooth. I have said that dentistry does not give social position, neither does it wealth. In full practice a dentist may get two or three cases in a month, and, for some, he may receive as high as five dollars; but that is a price far above the ability of the majority to pay, from one to two dollars being the usual rate.

"The base is always of wood. On the cheaper sorts the teeth are merely outlined upon the base, but generally they consist of ivory, shark's teeth, or stone, let into the wood, and retained in position by being strung on a thread, which is secured at each end by a peg driven into the hole where it makes its exit from the base.

"Iron or copper tacks are driven into the ridge

to serve for masticating purposes, the unequal wear
of the wood and metal keeping up the desired
roughness of surface.

"To construct a full upper and lower denture
requires about two days' constant work. Generally,
however, four or five are taken, as there must be
time allowed for the usual smokes and occasional
naps which are considered so necessary. The
ordinary service of a denture is about five years,
but they frequently last much longer. The writer
has one in his possession that has been in use
fifteen years, and is still quite serviceable.

"The process of manufacture is crude in the
extreme. A piece of wax large enough to cover
the roof of the mouth is heated, introduced, and
pushed up in position by the thumbs; it is then
removed and placed in cold water to harden.
Another piece of wax, large enough to make the
model, is then heated and applied to the impression, pressed into every part by the fingers, then
chilled by placing in cold water, and separated. A
piece of wood is now roughly cut to the desired
form, and the model, having been smeared over with
a red paint (veni), is now applied to the plate; where
they touch each other is marked by the paint.
This is then cut away and the process repeated
until the plate coats uniformly; it is then tried in
the mouth and any necessary corrections made.
They do not seem to be very particular to get a

smooth surface, at times not removing the tool marks.

"Whether or not these upper or lower dentures can be worn alone I do not know, never having seen them other than double; nor have I seen a single one carved to antagonize with the natural.

"As the teeth are not natural in appearance, and don't add to the beauty of the wearer, they are never used for purely ornamental purposes. I will except, however, those partial dentures which are fastened to the adjoining teeth by a piece of thread, they being used only for appearance sake; from the mode of manufacture there is no reason why one denture might not be made to answer several persons in succession." *

Some few years ago a young Japanese† came to this country to study the art and science of dentistry. He brought with him an example of the plate-work of his own people. The plate was a rim of hard wood, skillfully carved, in which were set pointed pegs of steel. This was one of the sets of teeth which from time immemorial the dentist of this nation had constructed for the toothless, who still retained an appetite, though the machinery for its indulgence had disappeared. This curious form of a denture was used merely for

---

*Dental Cosmos, vol. XIV., pp. 5, 6, 7.
†Demorest's Family Magazine, vol. XXIX., p. 82.

mastication and was removed and scoured as soon
as the meal was concluded.

The Japanese are a very dexterous people, and
if superstitions could be eliminated much in a
mechanical sense might be expected.

Among them superstition cries loudly—protest-
ing at all attempts at frustrating the plans of Him
who takes but what He gives.   "It were wicked,"
they claim, "to attempt to 'restore' what our God
has taken away."   Hence, with this belief deeply
imbedded, Dental Prosthesis can gain no strong
foothold. Prof. W. E. Griffis, of the Imperial
College at Tokio, Japan, says : " The Japanese are
admirable workers on metal. Iron, copper and brass
are wrought in every part of the country, and the
swords of Japan have long been famous. The
ornaments upon their hilts and handles, made of
copper, silver or gold, with inlaid work of various
metals, are not only curiosities but works of high
art.   They are most skillful in carving wood and
metal ; their lacquering in wood excels that of all
other nations." *

ADDITIONAL BIBLIOGRAPHY.

The Land of Art—Japan—M'Clachie ; Manners and Customs of Japanese—Mrs. Busk ; History of Medicine Among the Japanese—Wise ; Transactions of the Asiatic Society of Japan ; Glimpse at the Art of Japanese—Jarves.*

---

*See appendix for additional notes.

PHOENICIAN DENTAL ART.

Ancient Phoenicia, bordering on the eastern coast of the Mediterranean Sea, was particularly noted for its two great cities, Sidon and grand Tyre, and these cities in turn were famous for their manufacturers and artists. It was on this soil that King Solomon's temple stood, the grandest structure of antiquity. It was in the time of the Persian monarchy that Phoenician art reached its highest development, influencing all the nations around her. With Phoenician art and science the modern world has been little acquainted until the discoveries made by Gen. de Cesnola,* the results of which are in the Cesnola collection at the Metropolitan Museum, New York City. For three years this eminent Italian archaeological explorer employed several hundred men to excavate the ancient Phoenician cities, but principally Cyprus. During this period upward of 8,000 tombs had been opened, and a magnificent collection of antiquities gathered. It is the largest archaeological collection in the world and contains some 13,000 articles, mostly of precious metals. The Phoenicians, we learn, adopted from the Egyptians the custom of depositing their dead

---

*American Cyclopaedia, vol. XIII., p. 779.
  Ibid, vol. IV., p. 215.
  Encyclopaedia Britannica., Am. Sup., vol. II., p. 138.

in sarcophagi.* The oldest examples of anthropoid
stone coffins are made after the pattern of the
Egyptian mummy cases. Modern excavations
show that beside stone coffins, the Phoenicians
employed coffins of wood, clay and lead, to which
were often attached metal plates. Embalming also
seemed to have been frequently practiced as well
as covering the body with stucco. Great care was
bestowed by the Phoenicians on their burial places,
and their cemeteries are the most important monuments left to the post-civilization. The tombs
are subterranean chambers of the most varied form.
The dead were deposited either on the floor of the
chamber in the sarcophagi, or, according to later
custom, in niches. The mouths of the tomb were
securely walled up and covered carefully. Thus
safely stored away from time's ravishing elements,
the Phoenician race is preserved for modern research
and study. Archaeological societies have begun in
earnest to disclose these buried treasures, and the
dental profession already have received several
encouraging mementoes.

A specimen of ancient Phoenician dentistry is
accurately described by M. Ernest Renan, in his
work entitled "Mission de Phoenicie e le Cam-
pangne de Sidon," † as follows :

*Encyclopaedia Britannica, vol. XVIII., p. 810.
Anderson's General History, pp. 62, 63.
Museum of Antiquity—Haines, pp. 643, 911.
McCabe's Pictorial History of the World, pp. 180, 181.
†Page of the vol., 472.

"But that which was most interesting was the upper portion of a woman's jaw showing the two superior cuspids and four incisors, united by gold thread.   Two of these incisors seemed to have belonged to another person, and to have been placed there in order to replace the missing one.   This piece, which was found in one of the most ancient vaults, proves that the art of dentistry was pretty far advanced at Sidon, and also proves that the earth scurvy (scorbot de terre) so commonly seen nowadays in Sidon existed already in those ancient times." *

This piece of Phoenician dentistry, Dr. Van Marter informs us, can be seen in the museum of the Louvre, Paris.   There are scores of specimens of Phoenician dental art in home collections and also at the Columbian World's Fair.

Dr. Eames† assures us that "One of the earliest records of a dental operation is found upon a Scythian vase, discovered in an immense tumulus or buried mound, situated about four miles to the westward of Kertch, a small town on the Crimean peninsula.   Historically we know but little of the Scythians, beyond the meagre facts recorded by Herodotus, but in the almost numberless tumuli which are found upon the Crimean coast are pre-

*Independent Practitioner, vol., VII. pp. 58, 59.
J. G. Van Marter, D. D. S., Rome, 1886.
†Scythian Dental Art—Dr. W. H. Eames, 1886.
Independent Practitioner, vol. VII., p. 290.

served a most graphic record of their daily lives, manners and customs, in the funeral vases and other objects deposited in the final resting places of their dead.

"The richest of the numberless tumuli so far opened is one called the Koul-Oba, which was examined under the direction of the Russian government, and, although the greatest care was taken to preserve the precious relics, the larger parts were stolen and never recovered. The Koul-Oba was a royal tomb, and in a spacious apartment constructed of large blocks of stone was found the mouldering remains of the king, his queen or favorite wife, his servants, horses, and surrounded by his treasures. Near the splendid wooden sarcophagus of the king were the remains of a woman, doubtless his queen. On her head was a mitre-shaped diadem, and at her feet a vase of electrum, upon which is embossed a frieze of characteristic episodes of Scythian life. Upon the vase are four groups in exquisite repoussé work, giving incidents in their life. The King is clad in a Scythian costume, a tunic belted at the waist, and full trousers tucked in the boots; in one group he is listening to a report of a warrior kneeling before him; in another he is bending a bow, in a third his wounded leg is being dressed, and the last, as before stated, is one of the earliest known representations of a dental operation. The King is half sitting, half

kneeling, and the Scythian dentist is extracting a
tooth from the left side of the jaw.   It is reasonable to suppose that this represents an actual inci-
dent in the life of the king found in the tomb, for
in his skull, now deposited in the museum at
Kertch, the first and second left lower molars are
missing and the third molar is badly decayed. The
presence of an alveolar abscess connected with
these lost teeth, at some period of life is shown by
the condition of the alveolar process in this region.
"The only clue to the identity of the powerful
monarch here entombed is an inscription of three
letters upon an ornament, in which it is claimed
can be recognized the initials of a Bosphorus King,
Pairisades, who reigned about 310 B. C."

The vase, of which Dr. Eames speaks, was no
doubt Phoenician art, for it was about at this same
period, 310 B. C., that Phoenician pottery* claimed
the attention of the then ancient world. These
people were wonderfully skillful in ceramic work
and it is highly probable that the Scythian King
engaged his Phoenician neighbor to create the vase
and other earthen figures which modern research
has found within his tomb.

J. H. Middleton, an authority on pottery says :"
"Excavations in Cyprus, Rhodes, Thera, Troy,

*American Cyclopaedia, vol. XIII., p. 779.
Encyclopaedia Britannica, vol. XIX., p. 605
Peoples' Encyclopaedia, vol. III., p. 1413.

Attica and the coast of Italy have revealed the existence of an abundant class of pottery of great antiquity, a large part of which in its forms and decorations appears to have been Phoenician."* We learn that the Scythians† were not familiar with porcelain and clay work, and this account further proves that the vase containing the portrayal of a dental operation was of Phoenician origin.

ADDITIONAL BIBLIOGRAPHY.

Ancients and Their Teeth—Duval; Historical Research Into the Nations of Antiquity—Heeren; History of Civilization—Augsbury; Phoenician Inscriptions at Cyprus—De Vogue; Die Phoenizier—Movers; History of the Ancient Orient—Maspero; and Phoenician Arts or Mission de Phoenici—Renan.‡

---

*History of Pottery—Middleton, 1889.
†Herodotus, vol. IV., pp. 81, 97, 142.
 Hippocrates, vol., II., pp. 66, 82.
 Encyclopaedia Britannica, vol. XXI., p. 576.
 American Cyclopaedia, vol. XIV., p. 726.
‡See appendix for additional notes.

ARABIAN DENTAL ART.

The rise of the Mohammedan empire which influenced Europe so deeply, both politically and intellectually, made its mark also in the history of medicine and surgery.   After the Arabian conquest had ceased and the people consolidated, learning began to flourish ; schools of medicine,* surgery and pharmacy were established often in connection with hospitals and colleges, much after the fashion of to-day.

Although the Arab thought more of his steed than of his wife, yet he did not fail to appreciate self and give time and attention to the " pillars of the mouth," as he called the teeth.   Among the archives of tradition, in Arabia, we are informed that the augur and physician, Navius Aetius,† as early as 300 A. D., discovered the foramina in the roots through which the nerves and vessels enter the pulp chamber ; and for years subsequent to this discovery the christian world was ignorant of this and other of his important finds.

*Encyclopaedia Britannica, vol. XV., p. 805.
Peoples' Encyclopaedia, vol. II., p. 1133.
Encyclopaedia Britannica, vol. XXII., p. 675.
American Cyclopaedia, vol. XI., p. 348.
Dental Cosmos, vol. XII., p. 160.
†Dental and Oral Science—Dexter, p. 3.
Encyclopaedia Britannica, vol. I., p. 224.

Arabians never cease boasting of Aetius, who, at one time, was a professor * and tutor in the medical and dental departments of the celebrated Alexandrian University. He lived in the fifth century and wrote extensively on medicine, surgery and dentistry,† his medical writings filling some sixteen volumes. He, it is claimed, gave the first correct anatomical description of the teeth relative to their being supplied with nerves from the trifacial ;‡ and he also recommended the filling of decayed teeth with resinous substances, such as wax galbanum.§ He is also the first, as far as historical records lend assistance, to advocate the use of the file in dental operations. His advice to file the teeth freely to remedy their irregularity, was, however, founded in error, and would not meet with much favor among the skillful modern dentists.

Another Arabian surgeon, Albacasis ‖ by name, was recognized to be a great and preeminent "carver of human teeth." This famous doctor lived about 1100 A. D., and he suggested means for replacing lost dental organs by substituting extracted natural ones ; he also produced many dentures of bone and ivory.

The Arabs generally were very proud of their

---

*Dental Cosmos, vol. X., pp. 346 and 349.
†Professional Etchings—Dr. Kingsbury, 1868, p. 2.
‡Dental Cosmos, vol. X., p. 346.
§Ibid.
‖Dental and Oral Science—Dexter, p. 3.

teeth and among them the toothpick was a pocket friend.   The toothpicks were made of mastic wood and they used them in preference to quills ; hence Rabelais says that "Prince Gargantua, like the Arabians, picks his teeth with mastic-wood tooth pickers."*

They fully believe in their ancient Arabian adage: "He who does not masticate well is an enemy to his own life."†

Dr. J. R. Duval, in writing of the care which Arabians give their teeth, says: "The custom of washing the mouth every morning, which is adopted by several nations, has become the subject of a religious precept among the Arabians to make the little ablution,"‡ says Turnefort, "with the face turned toward Mecca, they rinse the mouth thrice and clean their teeth with a brush." This custom shows how highly the preservation of the teeth is esteemed by a people who formerly were forbidden, according to Menavius, "to have a tooth extracted without permission from the chief."§

In a recent number of the Harper's we read that a traveler who had visited and acquainted himself with the fleet Arabians, says: "If you are ever intending to visit the land of the 'ship of the

---

*Readers' Handbook—Brewer, p. 1017.
†Youth's Dentist—Duval, p. 13.
‡Voyages to the Levants—Turnefort.
§Youth's Dentist—Duval, p. 46.

sand' do not fail to comment the Arabian on his grand steed and pearly teeth ; these are the two things that are cherished in his heart."

But notwithstanding that the Arab never neglected his teeth or steed, we find that some of his good people wore artificial dentures ; at any rate an Arabian general under Mohammed of Ghor, and conqueror of India, was slain and his body could only be identified by means of the false teeth he wore, and held in place by gold wedges and wire.*

Arabians are not noted for their superstitious beliefs, yet they, like most of the ancient peoples, have strange ideas and conceptions relative to influences on humanity. If, in the event of battle, a soldier lose a tooth or teeth, he is not, according to the Koran, permitted to wear in their stead artificial substitutes, since the great law distinctly states "That he who fought for his country must fight even to the losing of his teeth."†

The great African explorer, Paul B. DuChaillu, states that in Arabia and Abyssinia can be seen people with the incisor teeth of both jaws filled to a sharp point in order to give a stern and savage aspect to the countenance, and also lend, as they claim, additional "beauty to the expression." The

---

*Johnston's Dental Miscellany, vol. VIII., pp. 80, 81.
  Rise and Fall of Mohammedan Power In India—Briggs.
†Dental Cosmos, vol. X., p. 17.
  Professional Etchings—Kingsbury, 1865.

explorer adds that "No female was entitled to marked consideration, or be known as a belle, without this peculiarity of the dental organs."*

Rhazes, and also Ebu Sina, both Arabian medical men, made use of white arsenic to devitalize the pulps of badly decayed teeth. The only aim seems to have been to produce a painless decay of the tooth.

The great historian, Wuestenfeld, tells us that upwards of three hundred Arabian physicians have left voluminous writings in Arabic, and not more than four works of this number have been translated, and these few exist in the Latin only.† Thus the great remainder, still in the form of Arabic manuscript, can lend us no aid in determining the exact status of the medical profession. "But it is improbable," says Dr. J. F. Payne, "that further research will alter the general estimate of the value of Arabian medicine. There can be no doubt that it was in the main Grecian medicine modified to suit the climates, habits and national tastes, and with some important additions from oriental sources. The Grecian part is taken from Hippocrates, Galen, Discorides and later Greek writers."‡ This being the case we can indirectly get a fair idea of

<hr>

*Dental Cosmos, vol. X., p. 346.
 Youth's Dentist—Duval, p. 41.
 †History of Medicine—Payne.
 ‡Encyclopaedia Britannica, vol. XV., p. 506.

Arabian dental progress by studying the Grecian dental art.

ADDITIONAL BIBLIOGRAPHY.

History of Arabian Medicine—Leclerc ; The Medicine—Celsus ; History of Medicine—Aegineta ; Mechanical Inventions by Arabs—Palgrave ; Arabian Philosophy, Etc.—Wallace ; History of Progress Among the Arabs—Palgrave. *

---

*See Appendix for additional notes.

GRECIAN DENTAL ART.

"The Greeks," it is said, "learned what the Egyptian's knew," and no doubt the science of dental surgery emigrated from Egypt to Greece as did nearly all knowledge. The Greeks, who were at one time a very dexterous and progressive race, were skilled in all arts and science,* and to their ancient historians and bards belong much credit for having noted the then present state of Dental Prosthesis, not only in the land of the Greeks but the land of man.

Homer, the great Greek sage and historian, tells us that Æsculapius, a surgeon who lived about 1250 B. C., used a narcotic to produce insensibility when performing minor operations such as tooth drawing. He, too, we are informed, was the first to teach the art of tooth purging and filling. Æsculapius performed many wonderful operations and his skill, it is claimed, enabled him to cure the most desperate diseases. He was thought of so highly by the Greeks that a statue of him was made in gold, and after his death was called the "god of medicine."

*Any Unabridged Encyclopaedia.

But Æscualpius * is not now recognized as the father of medical science, but only of medical practice—a distinction with a difference. Under his reign diseases were conceived to be "emanations from the anger of the gods," and cure was to be effected largely by their propitiation. A strict regimen was rigorously inculcated ; the temples were situated in salubrious places, and all that art could do toward stimulating the imagination of the patient was accomplished ; but the more careful study of the symptoms and causes of disease was unknown. Many changes in medical practice have taken place since then, but it is not certain that these changes represents improvements. There is, indeed, abundant reason for asserting that the hygienic injunctions of Ancient Greece were more nearly in accordance with the teachings of nature than those of our present medical professors.†

"The greatest surgeon that ever lived," says Herodotus, "was Hippocrates, who lived about 450 B. C. This genius was a distant relative of Æsculapius, and like this great surgeon was divinely skilled in the practice of medicine and surgery."

---

*American Cyclopaedia, vol. I., p. 152.
Encyclopaedia Britannica, vol. I., p. 209.
Peoples' Encyclopaedia, vol. I., p. 34.
Zell's Encyclopaedia, vol. I., p. 33.
†Dental and Oral Science—Dexter, p. 2.
Is Medicine a Science—Walker, 1886.
Phrenological Journal, vol. XXXIII., p. 285.
Encyclopaedias mentioned in the preceding note.
Dental Mirror, vol. I., p. 69.
Dental and Surgical Microcosm, vol. II., p. 2.

At the time of the birth of Hippocrates * medicine and surgery was entirely in the hands of the heathen priesthood, who knew little of medicine as a science, and so thoroughly clothed the subject with superstition and mysteries that future generations still suffer the effect. Every minute operation which these priests performed was accompanied by some special religious ceremonies. Temples were erected, and within their walls mythological figures stationed, each and every god or goddess being recognized as the divine guard against some disease, accident, pain or ailment. These priests † who had these several temples reaped at their doors unknown harvest of riches ; since if some poor unfortunate plebian suffered from toothache or other malady he would, by payment of toll or tariff, be permitted to enter the temple and there at the feet of the specially created god or goddess patiently pray until the pain subsided ; if it happened to continue for more than a day the wretched sufferers would return to the temple the next day with less money, less hope and more pain.

---

*American Cyclopaedia, vol. VIII., p. 740.
Encyclopaedia Britannica, vol. XI., p. 852.
Peoples' Encyclopaedia, vol. II., p. 907.
Zell's Encyclopaedia, vol. I., p. 1148.
†Dental Review, vol. III., pp. 429, 430, 431.
First Period in History of Dentistry—Patrick, 1889.
History of Medicine—Payne, p. 3.
Encyclopaedia Britannica, vol. XV., pp. 799-805.
American Cyclopaedia, vol. XI., pp. 345, 346, 347.
Past, Present and Future of Anaesthesia—B. J. Cigrand, 1890.

But, alas, these shrewd monks soon lost the substantial portion of their income when the young Hippocrates grew up "learned in all the wisdom of medical science and surgical art." Although himself the son of a priest-physician * and inheriting all the superstition, and educated in the traditions of the priestly rites, he broke loose from former teachings and proclaimed to all the civilized world that medicine was based on inductive philosophy, and disclosed at the risk of his life that the priestly system was a fraud and an imposition. He it was who first † undertook to collect the fragmentary knowledge of medicine and restore it to something of an order. He classified and described diseases, and with him medicine and surgery began their careers as sciences. All that is known concerning early history of medicine and surgery is derived from the works of Hippocrates, his family or his pupils. When we consider the age in which he lived—400 B. C.—and the difficulties under which he studied medicine, we can not fail to admire the great advance he made. His system is conspicuous in rejecting the superstitions of his time by teaching his many disciples to impute a proper agency to physical cause. It was to the interest of those connected with the temples to refer all diseases to supernatural agencies, and any contra-

---

*Any Unabridged Encyclopaedia.
†Encyclopaedic references on Hippocrates as already given.

diction of such doctrine by Hippocrates must have met with great reprehension. Yet the opposition seemed to weigh but little with this great and talented physician. He pursued his practice without giving himself the least concern in that respect, and in doing so set an example to all who should succeed him in his noble profession, and most forcibly taught his pupils not to hesitate in encountering the prejudices and superstitions of the present, for the sake of truth in the future.

We, as dentists, should reverence the memory of Hippocrates for the complete manner in which he accomplished his object. While Hippocrates investigated all branches of medicine, the diseases of the mouth and teeth did not escape his notice. He observed the teeth* in their healthy and diseased condition from the time of their appearance until lost in old age. Some of his quotations are as follows: "Teeth in similar conditions will erupt with less difficulty in winter than at any other period ; that children who sleep well have little difficulty in erupting their teeth ; that teething children with open bowels are less subject to convulsions than those suffering from constipation." He directed the attention to the influences "that diseased teeth have in diseases of the breast, throat and ears, which diseases he claimed could only be

---

*He wrote on the teeth in 460 B. C.

cured by removing the troubling teeth."* He further remarked: "Cold drinks affect and injure the teeth ; different seasons of the year have their various influences on the dental organs ; ulceration and suppuration of the gums can be prevented if proper care be given the teeth ; loose teeth" he advises to be "tied to their neighbors by means of gold or silk thread ; cleansing the teeth is a precautionary measure against decay ; avoid eating hard substances, thus not break or crack the teeth ; substances that set the teeth on edge are injurious." In one of his note books occurs this observation: "In consequence of a diseased tooth the maxillary bone of the son of Metrodorous became disorganized, the gums grew exuberantly, but the suppuration was moderate. He lost the molars and the maxillary bone."†

Those who studied general or dental surgery under Hippocrates were obliged to subscribe to what is known as the Hippocratic oath,‡ which was as follows: "I swear by Appollo, the physician, by Æsculapius, by Hygeia and Panaca, and by all the gods and goddesses, that to the best of my power and judgment I will faithfully observe

*Dental Cosmos, vol. XXIII., p. 670.
Dental and Oral Science—Dexter, p. 2.
Works of Hippocrates.
Youth's Dentist—Duval, p. 3,
†Dental Review, vol. III., pp. 433, 434, 435.
First Period in History of Dentistry—Patrick, 1889.
‡Archives of Dentistry, vol. I., p. 370.

this oath and obligation. The master who has instructed me in this art I will esteem as my parents, and supply, as occasion may require, with the comforts and necessaries of life. His children I will regard as my own brothers, and if they desire to learn I will instruct them in the same art without any reward or obligation. The precepts, the explanations, and whatever else belongs to the art, I will communicate to my own children, to the children of my master, to such other pupils as have subscribed to the physician's and surgeon's oath, and to no other persons. My patients shall be treated to the best of my power and judgment, in the most salutary manner, without any injury or violence ; neither will I be prevailed upon by another to administer pernicious physics."

After the death of Hippocrates the science of medicine and surgery took a retrograde step and again fell to the hands of the pagan priests, who made use of the discoveries of Hippocrates to further the confidence that divine power was invested in the priesthood.

"Such was the condition of medical observation in the then enlightened Greece," says Dunglinson, "confined to the priesthood, and full of mystery to the uninitiated ; but leading to a know-

---

*Works of Hippocrates.
Archives of Dentistry, vol. I., p. 379.
†Is Medicine a Science—Walker.
Phrenological Journal, vol. XXXIII., N. S., p. 284.

ledge of numerous remedial agencies, such as helle-
bore, opium, squill, blood-letting, etc.," and
"where sensible agencies failed, recourse was had
to charms, incantations and amulets, suggested by
ignorance and superstition among the rude and
barbarous nations of the present day almost as
extensively, and confided in as implicitly as in the
cradle of mankind. If the patient died the event
was ascribed to the will of the gods; if he recovered—by virtue of those instinctive powers which
are seated in every organized body, and without
which the efforts of the physician would be vain—
a case of cure was recorded, but no inquiry was
made as to the precise agency exerted. To the
charm, the incantation, the amulet was ascribed
the whole result. Tradition handed down a knowledge of its presumed efficacy, and led to its
employment in similar cases."

The century after the death of Hippocrates is
a time almost blank in medical science. Though
none of Aristotle's writings are strictly medical,
his researches in anatomy and physiology con-
tribute greatly to the progress of medicine and
surgery.* It should also be remembered that he
was of the Æsculapae family and received that
patriotic medical education which was traditional
in such families; though having practiced medicine

*Book of Problems—Aristotle, p. 229.
American Encyclopaedia, vol. XI., p. 346.
Encyclopaedia Britannica, vol. II., p. 518.

and surgery but a short period, his observations
were numerous.   It is probable that medical
science,* like others, shared in the general intellectual decline of Greece after the Macedonian
supremacy.   But Grecian sciences were revived
and the intellectual activity restored by the conquests of Alexander the Great.   More than one
learned center, in which medicine and surgery,
among other sciences, was represented, was instituted.   Pergamus† was early distinguished for its
medical schools, but in this, as in many other
respects, its reputation was ultimately effaced by
the more brilliant fame of Alexandria, where
Greeks taught and learned.   Here the eminent
Grecian surgeons, Herophilus‡ and Erasistratus,
practiced and thus taught the world the art and
science of medicine.   It was Erasistratus§ who
gave his particular attention to the dental organs,
and there is good grounds to believe that he was
an instructor in the dental department of the Alexandrian University.   If we can place reliance in
the words of the Greek philosopher, Aristotle, we
must believe that it was Erasistratus who deposited

----

*The Medical Schools of Ancient Greece were: Cnidos, Cos, Rhodes
Cyrene, Croton and Pergamus.
  †American Encyclopaedia, vol. XIII., p. 291.
  Encyclopaedia Britannica, vol. XVIII., p. 527.
  Peoples' Encyclopaedia, vol. II., p. 1352.
  ‡Dental and Oral Science—Dexter, p. 2.
  Any Unabridged Encyclopaedia.
  §Encyclopaedia Britannica, vol. XV., p. 801.
  Peoples' Encyclopaedia, vol. I., p. 661.

a pair of leaden forceps in the Temple of Delphi. The leaden instrument bore the Greek name "Odontagogos,"* meaning "tooth-extractor."

Upon the death of this great surgeon, and also on the destruction of the Alexandrian library, the medical art retrograded, and to this day Greece has not recovered from the shock.

The Greek writer, Cicero, gives credit to the third son of Æsculapius† for the invention of an instrument for the extraction of diseased teeth. It is further claimed that these ancients were acquainted with the art of healing dental caries by plugging the cavity with gold foil, and that the British museum contains skulls taken from Greek tombs, and the teeth of these aged remains are unmistakably filled with gold foil.

In a museum at Athens, in an upper jaw of an ancient cranium, a tooth filled with pure gold foil‡ can be seen. The skull was found in one of the tombs which date from the days of ancient Greece. From this it will appear that these ancient people, too, were familiar with gold work.

Dr. Xavier Landerer, of Athens, sends the

---

*Dental Cosmos, vol. VIII., p. 670.
Works of C. Aurelius—Medicine.
Dental and Oral Science--Dexter, p. 2.
†Dental Cosmos, vol. XXIII., p. 670.
Vierteljahrsschrift des Vereins deutcher Zahnkuenstler, 1881.
‡Dental Cosmos, vol. XXIII., pp. 109, 110.
Dental Cosmos, vol. VIII., p. 607.
British Journal of Dental Science, March, 1867.
Some Remarks On the Prevalence of Dental Caries Among the Ancients—James Bate, 1867

following to the London Chemist and Druggist: "It may be safely asserted that the ancients certainly cleaned their teeth and used tooth-powder. If the necessary attention were given, relics would be found in the graves of the women. The word 'odontotrimma,' the tooth-scouring stuff or tooth-powder, is found in ancient Greek, and in the Greek Pharmacopoeia is applied to tooth-powder. It is interesting to find that the ancients had made some advance in dentistry. A friend of mine who occupied himself in collecting ancient Hellenic skulls, wishing to show that they did not differ in shape from those now carried in Greece. Among several hundred of these skulls, some perhaps two thousand years old, we found two with 'stopped' teeth. One was filled with a mass as hard as stone, which, on analysis, proved to be hydraulic lime, made from volcanic ash, Santorin earth and lime. Marvelous as it may seem, the hollow of one tooth in the other skull had been filled with gold thread or gold-leaf. The metal used was pure. The skull itself, though deprived of the stopping, is now in the Archaeological museum."

The following brief quotation from a dental journal also portrays a few dental items:*

"The Greeks† appear to have been the first

---

*Dental Cosmos.
  The Pharmacist—1881.
  †New York Dental Journal, vol. II., p. 2.
  Dental and Oral Science—Dexter, p. 2.

who made any distinct progress in theoretical dentistry, although in many principles were wrong and their practice cruel and unnecessary, it was still something to find dentistry reduced to what at least approximated to a science. They seldom extracted a tooth except with the fingers. We also learn from Galen that various compositions existed among the Greeks and Romans for cleansing the teeth similar to our own modern dentifrices. Thus Andromache, Aristocrates, Crito and Diocletian were noted as amateur manufacturers of tooth-powder."

In the works of the renowned archaeologist, Belozoni, we read that "the Greeks wore false teeth of sycamore wood which had been fastened to the adjoining natural ones by ligatures of gold or silver ; and that many of the decayed natural ones were filled with a clay-like substance which became remarkably hard and durable."

The dead were either burned or buried, and in the latter case graves, vaults and tombs were used for the final deposition of the body.* For the burning of the body, piles of wood, called purae, were used, and oils and perfumes were thrown into the flames. When the pyre had burned down, the

*The tenth of the celebrated Greek Laws of the Twelve Tables (relating to funeral ceremonies) has, among others, this direction: "Let no gold be used, but if any one has his teeth fastened with gold, let it b lawful to bury or burn that gold with the body."

Dental and Oral Science—Dexter, p. 2.

remains were extinguished with wine, and the
bones were collected, washed with oil and wine,
and placed in clay boxes or an urn. The male-
factors, traitors, and all other disqualified citizens
were denied burial, which was considered the highest possible dishonor. Very few of their distinguished dead were embalmed ; the greater majority
were consigned to the urn. Thus the Grecian
custom of cremating* the dead has caused the
scarcity of dental specimens. What proof we do
have of their knowledge of Dental Prosthesis is
found in the literature of these people. What has
been said of the accomplishments of the Egyptians
relative to the art and science of dentistry is quite
applicable to the Grecians who were intimately
acquainted with the attainments of the Egyptians,
and modeled their customs and habits much after
their ancient neighbors.†

How far this people in their wonderful civilization advanced in the art of dentistry‡ we may never
know, but from the same authority we are led to
believe that the men who practiced this branch of
the healing art were of equal education and polish
with those found in the other branches of medicine;

*Civilization among the Greeks—Mrs. Dorman Steel, p. 33.
Brief History of Greece—Barnes, pp. 78, 79.
American Cyclopaedia, vol. III., p. 452.
Encyclopaedia Britannica, vol. VI., p. 565.
†Any Unabridged Encyclopaedia.
‡Journal of British Dental Association of Dental Science. vol. V.,
p. 407.
American Journal of Dental Science. vol. XIV., p. 283.

for learning and scientific knowledge in all its various departments were taught and practiced by the priesthood ; and in order to obtain this knowledge the individual had to take the vows of the priest while young, and be reared in the order.

ADDITIONAL BIBLIOGRAPHY.

Greece and Greek Antiquities—Green ; Museum of Antiquity — Haines ; Ancient History and Archaeology of Greece—Grote, 12 vols., 1846 ; Grecian Archaeology—Pantazis, Pittakis and Lambros ; Rise and Fall of Athens—Bulwer ; Life and Manners of the Greeks—Guhl ; Life of Alexander the Great and His Times—Williams ; Manners and Customs of Ancient Greece—St. John ; Antiquities of Athens—Stewart ; Medical and Surgical Art Among the Grecians—Creighton ; Geschichte der Medizin—Haeser ; History of Medicine—Paulus Ægineta ; Life and works of Hippocrates—Greenhill.*

---

*See appendix for additional notes.

ROMAN DENTAL ART.

We will now bid the scholarly Grecians good
bye and travel to the land of war—Italy—there to
greet the model Roman.   Although the Romans
were constantly engaged in battle, yet some attention was paid to the finer arts.

In the early history of Rome the arts and
sciences made but little progress, and what advancements that came, were derived from the Greeks
and Egyptians.   We learn from Pliney "that the
art and science of medicine and surgery were introduced into Rome at a later period than the various
other human accomplishments."   He further writes
that "for nearly six hundred years the Roman
people were without medical or surgical aid ; not
that no attempt was made to cure disease, but that
these attempts * consisted mainly in superstitious
observations."   Thus, according to Levy, "following
the advice of the Sibylline books, pestilences were
repeatedly stayed at Rome by erecting a temple to
some favorite god."   Like their immediate neighbors, the Grecians, after whom they copied, made
priests of the temples custodians of divine cure.

In those ancient times to endow a monastery,

*Encyclopaedia Britannica, vol. XV., p. 802.

found one, or to have performed a miracle was the safest passport to canonization. The following, taken from an ancient work on mythological beliefs, gives a complete list of such saints and gods as the plebians were obliged to give devotion, in case of ordinary dental troubles: "Saint Apollonia guarded against toothache; Saint Lucy guarded against sore tooth; Saint Anthony guarded against inflammation; Saint Germanus guarded against diseased erruption; Saint Marcus guarded against neuralgia; Saint Herbert guarded against poisoned teeth."*

The Roman priests also erected temples in memory of the great Grecian physician, Æsculapius, and worshiped him as a god of medicine. The cock was commonly sacrificed to his memory, but a peculiar breed of serpents † was the favorite votive. The monks shrewdly worked upon the confidence of the laity as regards the miraculous cures wrought by praying to Æsculapius, and the priests of the temples at once founded the society known as Æsculapae,‡ or children of Æsculapius, and the members of this association were the only regular physicians and surgeons of antiquity.

---

*Development of Religious Beliefs—Gould, 1870.
Mythology of Ancient Italy—Keightley, 1844.
Roman Mythology—Poeller, 1865.
Dental Review, vol. III., p. 430.
Miracles of the Roman Catholic Saints—Rev. O'Connor, 1893.
†American Cyclopaedia, vol. I., p. 152.
‡Ibid.

"The sale of Æsculapian snakes, or holy snakes,"* as they were called, was a source of revenue to the priests and physicians who lived about the temples. Thus religion, medicine and surgery were practiced together, and through the instrumentality of priest-physicians sacrifices and votive offerings of value served to enrich the medical temples and oppose the anger of the immortal gods and saints at the same time.   Charms, talismans and amulets were resorted to for individual ailments and to ward off diseases.

The priests in simple had what we would now-adays term a "corner on medical science," and they jealously guarded the votaries and fought every advance made by the plebians towards medical discoveries.

Pliney says: "Thus the priests became the recognized surgeons and they taught the science with many acult and mysterious ceremonies well calculated to impress the vulgar and to excite belief in their miraculous power."   So we find that the great science of medicine and surgery was confined to the ancient priestcraft, and they took good care to keep it well saturated with mysteries ; but they themselves were well posted and studied laboriously to gain knowledge in the promising new field.   The priests discovered numerous medical

*First Period in the History of Dentistry—Patrick, 1889.
American Cyclopaedia, vol. XI., p. 345.
Ibid, vol. I., pp. 152, 153.

agencies, but the outside world was kept ignorant of these priestly accomplishments. While the science was being cradled by these inquisitors the common surgeon was prohibited from practicing, under penalty of imprisonment or death.

Sumner has wisely said that "Vice and barbarism are inseparable companions of ignorance and superstition, and without knowledge there can be no sure progress." Hence we comprehend why there was a pause in the research and advancement in surgical science; the priests, having full sway and unlimited power, kept all information from the laity, and only a select few, who had the necessary influence and pecuniary circumstances, could learn of the new and wonderful discoveries in any of the prosthetic arts.

But, kind reader, through an agency entirely unknown to antiquity, knowledge of every kind has become general and permanent; it can no longer be confined to a select circle, or crushed by tyranny, nor be lost by neglect. The press, ever watchful with its one hundred eyes of Argus, and strong with more than a hundred arms of Briarious, not only guards all conquests of civilization but leads the way to further triumphs.

Among the voluminous writings of the Latin poets frequent reference is made to artificial teeth. The famous Martial, who lived in the first century B. C., says that a Roman dentist, "Cascellius, is in the habit of fastening, as well as extracting, the teeth."* To Lelius the same author says: "You are not ashamed to purchase teeth and hair,"† and adds that "the toothless mouth of Ægle was repaired with bone and ivory ;" also that "Galla, more refined, removed her artificial teeth during the night."‡ The immortal Horace, of the same century, cites the case of the sorceresses, Canidia and Sagana, running through the city and losing, "the one her false hair, the other her false teeth."

Another quotation reads: "Thais has black teeth, Lecania has white ones. Why is this? The former has her own teeth, the latter purchased ones."§ The testimony of Martial refers only to the dentistry of the first century. But there are proofs of an earlier period of this art in the Tenth of the laws of the Twelve Tables, which date as early as 450 B. C., made exception to burying or burning gold with the dead, "such gold as is found in the mouths of the deceased for the purpose of fastening the teeth together."||

An eminent English scholar adds: "Cicero,

* "Eximit aut reficit dentem, Cascellius, aegrum"—Martial.
  Dental Cosmos, vol. XXIX., p. 75.
† Practical Treatise on Crown and Bridge-work—Evans, p. 1.
  Dental Cosmos, vol. XXIX., p. 45.
‡ Practical Treatise on Crown and Bridge-work—Evans, p. 1.
§ Ruyers Medical Studies on Ancient Rome—September, 1881.
  North American Med. Chir. Review—1881.
  Dental Cosmos, vol. II., p. 188.
|| Dental and Oral Science—Dexter, p. 2.
  Dental Cosmos, vol. II., p. 188.

when speaking of a law passed to check the unnecessary expense of funerals, says: 'Neve aurum
addito,' etc.; that is, add no gold to the funeral
offerings, but whosoever has his teeth bound with
gold 'suevi aurodentes vincti,' let it be no evasion
of the law to bury or burn him without it."*

The same authority continues, saying: "Any
mechanical dentistry, or prosthetic dentistry as
our American friends prefer to call it, that was
practiced in ancient Rome, appears to have been
rather primitive, as the following from the poet
Martial shows:

  " 'Thou hast only three teeth, and these
   Are of box-wood, varnished over.
   Thou shouldst fear to laugh ;
   Weep always if thou art wise.'†

"So box-wood was not only used for 'volubile
buxum,' or boys spinning tops, of which Virgil
speaks, but in the reign of Domitian was carved
into the shape of teeth."‡

When teeth are loosened by a blow or other
causes Celsus says "fasten them with gold to
those that are firm."§

The following from the works of Duval bring
forth a latent ray of Roman dentistry: ‖

  "To be able to laugh without fear of showing the

---

*Journal of the British Dental Association, vol. V., p. 353.
†Ibid, vol. V., p. 354.
‡Ibid.
§Dental Cosmos, vol. XXI., pp. 184, 279.
‖Youths' Dentist—Duval, p. 100.

teeth which have been skillfully filed, and to masticate freely with those which have been stopped with gold, are incontestible proofs of our art. But what a satisfaction to be able to have one or more artificially replaced which have been lost; with what alacrity is this innocent stratagem embraced, in order to hide the disorders of the mouth! Without it how many mouths would be spoiled! It restores to the physiognomy a part of the graces which it had lost, and it would have rendered 'null and void' the effect of an ancient law among the Romans, entitled 'Cui Dens,' the object of which was to examine if any one who had lost a tooth was in the possession of perfect health."

"It was about the second century B. C.," says Pliney, "when the medical profession was introduced into Rome, and it was an importation from Greece."* The first Greek physician was Arcagathus,† who became a practitioner of medicine in Rome 218 B. C. Unfortunately for him, but no doubt well for the people, he had too much confidence in his remedies, his practice illustrating his confidence, and his patients died, and Rome was aroused to prohibit the practice of medicine by law.‡ For more than a hundred years Rome was consequently without a physician, but as "the once proud mis-

---

*Encyclopaedia Britannica, vol. XV., p. 802.
 Any Unabridged Encyclopaedia.
†Is Medicine a Science—Walker, 1889.
 Phrenological Journal, vol. XXXIII., p. 285.

tress of the world," grown lofty by her conquests, and rich in all the arts of wanton pleasure, she began to decay in the first requisits of a great empire, a vigorous manhood, the art of the physician became a recognized desideratum, and soon again the sects were almost as numerous as those of their rival nation.

We learn that in the year 60 A. D., Andromachus, a Roman physician, invented "theriac" for filling* the decayed teeth. It was a mixture supposed to possess great healing properties. Subsequently theriac was considered the great panacea, and was made with great solemnity in open market-places of Rome and Venice.

Of the few Latin medical authors, Celsus, the first native Roman physician, is the chief. He lived in the first century and wrote voluminous treatise on architecture, philosophy, rhetoric and medicine, on which latter departure his eminence rests. His book, "The Medicine," is a digest of what was known to the ancients on the subject, and portrays the great progress made in consequence of the labors of the anatomists at the Alexandrian library. Celsus treated of most of the great surgical operations, and the operations on the dental organs did not escape his notice.

If we may place reliance on written history

---

*Dental Cosmos, vol. XXIII., p. 670.

Vierteljahrsschrift des Vereins Deutscher Zahnkuenstler—1881.

then we are obliged to believe that it was Cornelius Celsus, the noted Roman surgeon of the time of Tiberius, who wrote upon the diseases of the teeth * and their treatment, and he is also credited with inviting or introducing the art of plugging teeth with gold foil.

Celsus, in the year 32 B. C., used the root-forceps, which bore the Greek name "Rhizagra,"† although according to the name it would seem not to have been invented by Celsus, but by some Greek. Following Celsus came the learned and eminent Claudius Galen, born, A. D., 130; he studied the art of medicine at various recognized medical schools, but learned the practice principally at the Alexandrian library. He practiced his profession at Rome. He, above all others, gathered the divergent and scattered threads of ancient medicine and established landmarks which Father Time of to-day must recognize. He was a man furnished with all the anatomical, medical and philosophical knowledge of his time.‡ He found the profession of his time split up into a number of sects, and medical science confounded under a multitude of dogmatic and superstitious systems. He appears to have made it his express duty to reform these

*Dental Cosmos, vol. XXI., pp. 184, 297.
  Dental and Oral Science—Dexter, 1876, p. 3.
†Dental Cosmos, vol. XXIII., p. 670.
‡Any Biography of Galen.
  Any Encyclopaedia.

evils, to reconcile scientific acquirements and practical skill, to bring back the unity of the profession, and wrest the practice from the hands of the clergy. He accomplished all he anticipated and he wrote an encyclopaedia of medicine, and for upwards of a score of centuries his authority reigned supreme.* His knowledge of the teeth was extensive and he described accurately many dental diseases.

Subsequent to the death of Celsus and Galen, medical and surgical art once more came to the hands of the priestcraft ; the pagan priest had now differentiated, and recognized as Roman Catholic priests. For generations upon generations the Roman clergy were the sole monarchs of the medical profession. The extraction of teeth and all other minor operations were directly in their domain, and the general practitioner suffered much after the same fashion as did his Greek colleague. At that early period it was a part, and an important one at that, of the curriculum of the theological student to learn and practice the art of medicine and surgery. Fortunately for the medical profession, after repeated struggles the council of Tours,† held in the city of Tours, France, early in 1163, an edict was enacted in which the clergy were interdicted from all surgical practice. The clergy, how-

---

*Encyclopaedia Britannica, vol. XV., pp. 802, 803.
†American Cyclopaedia, vol. XI., p. 349.
Encyclopaedia Britannica, vol. XV., p. 806.

ever, were permitted to continue as physicians, and,
up to this late day of the nineteenth century, the
theological student of the Roman church is obliged
to acquire a fair knowledge of medicinal agencies,
and is taught their effective uses.   It was Guy-de-
Chauliac,* a Roman priest, who compiled from the
Grecian and Arabian authors the earliest works on
medicine and surgery.   He practiced the profession
at Lyons, France, and was afterwards employed by
three popes of Avigon, Clement VI., Innocent VI.
and Urban V.; he taught the art in one of the
ecclesiastical colleges of Rome, and was recognized
as a Roman authority on medicine.

A writer in one of our early dental journals†
tells the following:

"Until this period (middle ages) nothing had
been employed for filling the teeth but resinous
and aromatic substances, intended rather to calm
the pain than to strengthen the member.   The
first to speak of filling with gold-leaf is John Arcu-
laus, who filled teeth with this material.   Jean
De Vigo permitted abscesses of the gums to ripen,
then treating them with honey and Egyptian oint-
ment.   He used for caries, especially the molars,
the file and rasp, and filled teeth with gold-leaf

*Encyclopaedia Britannica, vol. XV., p. 806.
American Cyclopaedia, vol. IV., p. 344.
Grand Surgery, Joubert, 1592.
†New York Dental Journal, vol. II., p. 79.

apparently approximating to the modern treatment."

The Roman practice of cremating all but the most noted of their dead has consequently destroyed most of the desired evidence in this direction. Ovid and Virgil make similar remarks which prove, beyond the shadow of doubt, that Dental Prosthesis was a known art to ancient Romans.

As to the condition of the teeth in prehistoric and medieval or modern times I can quote no better authority than Dr. Talbot, who has devoted years, energy and finance to learn of the true condition of the dental organs or their substitutes of our historic ancestors. In an address * by the doctor (in 1891) he says, relative to the ancient dental practitioner: "The instruments for dental, as well as surgical purposes, which are to be seen in the museums of Europe, together with the beautiful specimens of Etrurian and Phoenician dentistry, now in the possession of Drs. Van Marter, of Rome; Barrett, of Buffalo, and Taft of Cincinnati, are striking illustrations of the superior ability which men of early times acquired."

These specimens of which Dr. Talbot speaks are rare, owing not to the supposed cause of dentures not being common in ancient days, but on account of the peculiar customs of these early folks in disposing of their dead, the few and treasured

*Dental Cosmos, vol. XXXIII., p. 456.

relics, which we as a profession possess, have come down to us simply because time, weather and circumstances did not destroy them.

In 1889 Dr. Barrett, while exhibiting some of the treasure-trove, remarked : * "These specimens date from about the founding of Rome. They are of more than unusual interest, as they bear unimpeachable testimony on some interesting points connected with the teeth of man. Dentists of to-day usually entertain the idea that the prevalence of diseases of the teeth is to be attributed to the altered methods of living, to the modes of cooking food, to change in the manner of life, etc. Some years since I had examined about two thousand (2,000) ancient skulls, more especially with reference to evidences of dental disease. The examination at once demonstrated conclusively that all the diseases of modern life, except syphilis, were as rife in ancient times as to-day. Two of the teeth I have shown you prove the existence of pyorrhea alveolaris in teeth seven hundred and fifty years before the Christian era."

Now, if this, as Dr. Barrett cites, is correct, which no doubt it is, there is abundant proof that in those days, like in our own times, there were, as Shakespeare says:

> "Sans teeth, sans eyes,
> Sans taste, sans everything."

---

*Dental Cosmos, vol. XXXI., p. 148.

And since Dr. Talbot comes to us with conscientious assurance that the ancient practitioners were highly skilled in the preserving, and as well, reconstructing dental organs, we must feel convinced that the old dentists practiced upon the "sans teeth," of which Dr. Barrett lends undoubted authority.

The reason why I dwell upon this subject of ancient dental substitutes is that we have modern scientists among our ranks who are attempting to popularize the idea that there are not in existence any authentic proofs of there having ever been in ancient times dental practitioners, nor that the mouldering dead demonstrate the science of crown bridgework or artificial teeth. Some years since one of our American dentists wrote an article for one of our dental journals, in which the author states that "no well authenticated case of gold filling has been found in the teeth of the ancient Romans, Etrurians or Egyptians, and that the superior cement reported to exist in the teeth of these dead has proved to be simply tartar." The same author adds that the socalled "bridgework" reported to have been found are crude attachments of artificial teeth by gold wire, and from large use in the mouth are thickly coated with calcareous deposits. Their mechanical contrivances are in no wise comparable with the artificial productions of to-day."

It gratifies me to tell you, as readers, that many genuine cases of filled teeth, crown-work and bridges which were exhumed in various parts of Italy, Greece, Etruria and Egypt, are on exhibition in the present Columbian World's Fair, and I hope the writer of the aforespoken disclaimer can have the pleasure of acknowledging his mistake.

In a recent copy of a scientific weekly paper I noticed the following seemingly wonderful disclosure: "It is claimed by the modern dental surgeon that ancient people had their teeth filled with gold, obviously to prevent further decay of the teeth. This, on recent and close investigation, proves to be, as the Americans term it, a fake, and it is easily proven to be such. What the anxious dentists of to-day thought to be gold plugs were clearified to be nothing more or less than mere gilding of the teeth. The belles of old were accustomed to lend aesthetic marks to their delicate features by gilding their front teeth."

This writer, too, no doubt, believes that he has solved an intricate problem, but in this case his little knowledge has proven to be a dangerous thing, since he does discredit to the ancients and attempts to deprive medical science of its laurels and simultaneously rob history of facts. Had the writer drank but deeper of the Pierian-spring he would have sipped up a draught of information such as would have not only cautioned him against writing as he did, but, on the contrary, stimulated him to regard records of the past as worthy of deep and continued study. It is true that in many of the tombs of ancient Italy and Greece the mummified corpses present golden teeth, or more clearly speaking, gilded teeth, and it will be of interest to us to learn why such was the case.

We learn in the Grecian as well as Roman mythology that in the event of the death of an eminent personage, and more especially the kings, emperors and public benefactors, the burial ceremony included the gilding of the teeth of the dead.* And why this you ask? For the simple and particular reason that their mythological belief proclaimed that the teeth be gilded, in order that the departed might greet the immortal gods of judgment in all possible glory. The river Styx† was by them supposed to be the boundary line between life and death, and in consequence the departed, in order to enter the kingdom of the immortal god—heaven —were obliged to cross this fearful river, Styx. But to facilitate the passage, the dead must be in the good graces of the god Charon,‡ the watchman of this stream and the ferryman of the shades of death,

*Roman Mythology—Haines, p. 485.
  Classical and Mythological Dictionary—Carleton.
  Independent Practitioner, vol. VI., p. 47.
†Any Unabridged Encyclopaedia.
‡American Cyclopaedia, vol. IV., p. 321.
  Encyclopaedia Britannica, vol. V., p. 430.

and in no way could the love of this god be secured other than by gilded teeth and copper coin. The fee exacted by him for this service from each spirit ferried over the Styx was never less than one obulus—one penny—nor more than three ; and to provide for this fee small coins were placed in the mouths of the dead.*   The spirits of those bodies which had not gilded teeth and sufficient copper coin to pass, were supposed to wander on the shore of the Styx for one century, after which period the god Charon would permit the unfortunates to enter the boat and cross the stream and be escorted to the seat of judgment.   Now then, kind listeners, we comprehend why some of the dead and mummified bodies of the ancients have gilded teeth.

I would caution any reader against believing too readily anything pertaining to the accomplishments of the ancient dentists; yet I also would advise him not to be too reluctant about yielding, especially when the facts and figures thoroughly demonstrate him as being opposite to truth and justice.

*Manners and Customs of Ancients—Mrs. Dorman Steele, p. 32.

ADDITIONAL BIBLIOGRAPHY.

Medicine and Surgery Among the Romans—
Payne; Christian Antiquities—Smith; The Papacy
—Mullinger; Roman Law—Muirhead, 1889; History
of Roman Law in the Middle Ages—Savigny, vol.
II.; Decline and Fall of the Roman Republic—
Long, 1845; Kunst und Kuenstler Italiens, 1878;
Antiquities of Rome—Cresy, 1821; Decline and
Fall of the Roman Empire—Gibbon, 1862; Hand-
buch der Roemisher Alterthuemer; Annals of
Roman Ecclesiastics—Baronius; Topography and
Archaeology of Rome—Middleton, 1889; Mythology
of Ancient Rome—Keightley, 1844; Manual of
Mythology—Murray, 1873; Records of the Past—
Schrader, 1874; Historic and Monumental Rome—
Hemans, 1872; Museum of Antiquity—Haines,
1885; Works of Galen—Cox; Medicine in Galen's
Time—Gasquet.*

*See appendix for **additional notes.**

ETRURIAN DENTAL ART.

The Etrurians, who inhabited the northern part of Italy, were well skilled in mechanical sciences, and Etruria flourished as the Italian seat of learning, wealth and power. These ancient Etrurians were a very remarkable people. Among them the fine arts were highly cultivated and dexterity so well developed as at the present time to excite admiration. They were exceedingly luxurious* in both dress and appetite, and extremely fond of personal ornaments, even going so far as to have their sound natural teeth gilded, a custom of which our modern belles cannot boast.

In those days the barber did not claim dentistry as a foster child, as the following depiction of a tonsorial shop, by Plutarch, about 73 A. D., will clearly show: "The barbershop, with its talkative inmates, was not only frequented by those requiring the service of the barber in cutting hair, shaving, cutting the nails and corns and tearing out small hairs, but was also a symposium house where politics and local news were discussed."† Had the barber in this ancient day practiced the science or art of dentistry, this writer, who was

---

*Any history of the Etrurians.
†Museum of Antiquity—Haines, p. 236.

very minute and exacting in his composition, would have made proper reference in the description.

Among the Etrurians dental science was studied and practiced as a specialty of medicine; "however, in this department of learning," says Professor G. A. F. VanRhyn, the eminent archaeologist, "the Etrurians were imitative rather than creative, and the art bore at every period the marks of foreign influence, especially Egyptian, Babylonian and Grecian."*

Notwithstanding that the Etrurian dentists patterned much after the oriental artists, yet much credit is due them for having perfected many dental operations of a more difficult character. We shall, in the near future, know more about the accomplishments of the Etrurians, since many archaeologists are hard at work solving the language of these grand people. They left us no key to their strange language, and no history except that which is written in the tombs, hence all we know of them is from adjacent and contemporary nations of people. Numerous are the theories advanced as regards derivation or origin of the Etrurian race. Simultaneous with the discovery of the key to the language shall come a long, interesting and profitable lesson relative to their attainments as dental practitioners.

---

*Independent Practitioner, vol. VI., p. 2.

The Etrurians,* like the Greeks and Romans, held great faith in the mythological gods, and much like the oriental people sought the good will of the angry immortals by prayers, votive offering and sacrifices.

Their priests, whom they called "lucumos," were the guards and guides of the various religious and medical temples, but the medical profession with its numerous branches was practiced with great success, as we learn from the object lessons left us.

Modern dentists feel somewhat flattered by their late success in crown and bridge work, but our prehistoric professional forefathers we find did the same ingenious work centuries ago.

In the museum at Cornets, Italy, can be found, carefully guarded with lock and key, two specimens of ancient Etrurian bridgework.† Their authenticity is undoubted since Van Marter, at present a Roman dentist, procured from the Sig. Dasti, the royal inspector of excavations and exhumations at Cornets, Etruria, a certificate duly signed and sealed, testifying that said specimens of gold bridge-work were discovered in the mouth of a corpse which had been entombed upwards of two thousand four hundred years ago. The cases

---

*Any Unabridged Encyclopaedia.
†Some Evidences of Prehistoric Dentistry in Italy—Van Marter, 1885.
Independent Practitioner, vol. VI., p. 243.
‡ Dental Cosmos, vol, XXXIII., p. 456.
Archives of Dentistry, vol. II., p. 87.

were well made, the artificial teeth were evidently carved from the teeth of some large animal, and were well executed. The artificial substitutes were the two superior central incisors and the first bicuspid of the left side. The artificial centerals and the natural lateral and cuspid of the right side were in a fair state of preservation, and the entirety was retained in position by gold bands, while the natural lateral, cuspid and second bicuspid and artificial first bicuspid of the left side were lost. The three substitutes were also encircled by gold bands secured by rivets passed through each tooth. Three cases more of a similar construction were unearthed in the crumbling Etrurian tombs.

These are the earliest known essays of dental bridgework. What conclusions are we to draw from the evidence of wonderful surgical instruments and appliances found in the ruins of Pompeii, instruments that have been re-invented in recent years to meet the demands of modern surgery? One is almost inclined to call a halt before expressing any opinion, and wait a little longer for excavators to dig up Etrurian or Urbain telephones and a long catalogue of similar supposed modern inventions.

Dr. Van Marter* contributes the following additional proof on Etrurian dental art: ''In the

---

*Dr. Van Marter deserves great credit for having produced proof of Etrurian dental art. To him belongs the credit of the greater portion of this section.

library of the Barberini Palace, Rome, most carefully guarded by lock, key and screw, I found this
specimen of early dentistry.   Viewed under glass,
this case might easily deceive the unprofessional
eye, for it was thickly covered with the debris of
ages.   It took a great deal of persuasion to induce
the polite and careful librarian to allow me to
remove enough of the dust of centuries to see what
the Etrurian relic really was.   It proved to be four
natural teeth, two superior central incisors, laterals
and cuspids, banded together with pure gold bands,
and attached to adjoining teeth   This case belongs
to the same period as those found at Cornets, and
in workmanship was so nearly identical that it
might have been made by the same dentist.   It
was taken from an Etrurian tomb at Palestrini,
near Rome, with numerous fine specimens of gold
and bronze work.

"The most recently opened and oldest Etrurian
tomb yet discovered in Italy, was lately excavated
at Capadimonti, near the lake of Bolsena.   The
entire contents of this tomb included three teeth
bound together with a band of pure gold, gold
spiral springs, silver finger-rings, necklace of amber
and glass, arm-band, bronze, vases, etc.   This
tomb belongs to the sixth century B. C.   There is
nothing to indicate that these three teeth were
attached to any adjoining teeth, and we are left to
conjecture whether they were loose natural teeth,

supported by the gold band around the lateral
and bicuspid.   It is not at all improbable that the
cuspid may have been a transplanted tooth, for
we are sure that in those early days they had very
clever surgeons, and slaves were made to serve
their lords and masters in any capacity, 'from
building grand temples and monuments to supplying teeth for transplantation.'   Certainly the spaces
between these teeth are wide enough to satisfy the
most rabid dental separatist, and the position of the
teeth does not indicate that perfect regularity and
symmetry were the invariable rule, even in those
early days.   This is significant, when we consider
that the former owner of these teeth was evidently
a lady of distinction, judging from the ornaments
and contents of the tomb.   At least this specimen
of early Etrurian dental work is of interest to us as
the oldest yet found in Italy, and as supplying one
of the missing links of the dental chain we are
endeavoring to trace back to the beginning of our
profession.   It is certain that dentistry must have
been extensively practiced in the early history of
the world and that gold must have been used
largely."

In 1884 the great English surgeon, Sir Spencer
Wells,* made investigations relative to Etrurian
medical and dental science and expressed great
interest in the matter; he related to Dr. Van Mar-

*Independent Practitioner, vol. VI., p. 4.

ter, D. D. S., of Rome, that he learned that these ancient folks had their teeth filled with a kind of fusible metal. The noted English archaeologist and writer, Mr. Forbes, while in Etruria and Rome, discovered that many of the mummified dead had teeth filled with gold* and a peculiar amalgam.

Hence we would conclude that few Etrurians suffered or died from toothache. On the contrary, we are inclined to think they had better teeth than we of nowadays have, and that in many respects these people were wiser than we.

As a rule they cremated their dead and this custom of theirs renders our task of procuring evidence a very difficult one. From what I can learn, only great warriors and civilians of distinction were embalmed and laid at rest in the family tomb. Two or three thousand years of time has accomplished the same end, since nearly all those who were embalmed and laid to rest in state suffered the same result, thus: "Ashes to ashes, dust to dust." This narrows our limits of research to a small territory, and makes it rather surprising that any symbol of dental work should come down to us from those remote times.

But in the days of flourishing Etruria, only the elite were fortunate enough to receive the benefits of dental operations, while the low and lowly were

---

*Independent Practitioner, vol. VI., p. 4.

forced to suffer the tortures which diseases of the teeth and oral cavity entailed.

The tombs in which the elite of those days were enshrined were most beautifully finished and their walls were ornamented with frescoes typical of the life at that period. Time has naturally covered these sacred vaults, and the ground above them has been cultivated for ages, while below are human ashes telling unknown legends of the lost art—Dentistry.

ADDITIONAL BIBLIOGRAPHY.

Antiquities of Etruria—Lanzi; Ancient Etruria—Mueller; Archaeological Discoveries in Etruria—Wachsmuth; Museum of Etrurian Urns and Sarcophagi—Brunn; Etrurian Jewelry and Metal Ornaments—Murray; Etrurian Arts and Sciences—Murray; History of Ancient Etruria—Deecke; Ancient Artists—Spence; Accomplishments of Ancient Etrurians—Steub.*

*See appendix for additional notes.

FALL AND REVIVAL OF DENTAL ART.

The science of dentistry from the fifth to the eighteenth century was entirely neglected, and to the suffering masses lost in oblivion during the long and blank period of human record, historically known as the middle ages. In this time the mere operation of extracting useless and painful teeth was the extent of dental science, thus this dark age not only retarded advancement in our science, but it produced retrogression; with but few occasional rays of light penetrating its misty veil, only to be immediately swallowed in the dense surrounding gloom of superstition and religious intolerance.*

During these dark days, known as the dark ages, all sciences and arts were completely neglected and the born artists, scientists and even the philosophers were, by cruel fate, turned into soldiers, knights and marshals. This gloomy period covered a duration of time estimated by historians to be about one thousand years.† All professions suffered in this reign of terror and the various callings of scientific men fell to the hands of mere artisans and laborers. Dentistry, once in

---

*Any complete History of the Middle Ages.
Any Unabridged Encyclopaedia.
†Any complete History of the World.

the hands of competent and deserving men, now took a retrograde step and became a branch of the blacksmith's, barber's and jeweler's trade. Oral surgery, or more properly speaking dental surgery, was then practiced by the barbers and blacksmiths exclusively. The following from George Elliot's Romola * (1492) well illustrates the tonsorial claim on the dental school: "Nay Bratti," said the barber in an undertone, "thy wisdom has much of the ass in it, as I told thee just now; especially about the ears. This stranger is a Greek, else I'm not the barber who has had the sole and exclusive shaving of the excellent Calcondila Demetrio and drawn more than one sorry tooth from his learned jaw."

In the same book is found † the following: "The Chirurgic Art!   *   *   *   Is it your Florentine fashion to put the masters of the science of medicine on a level with men who do carpentry on broken limbs, etc., and sew up wounds like tailors, and carve away excrescences as a butcher trims meat? 'Via!' A manual art, such as any artificer might learn, and which has been practiced by simple barbers like yourself—on a level with the noble science of Hippocrates, Galen and Avicenna, which penetrates into the occult influences of the

---

*Romola, p. 31.
†Ibid, p. 165.

stars and plants and gems—a science locked up from the vulgar?"

Medicine and surgery in all their branches, having early deviated from their true course, were soon given over to alchemy, necromancy and magic. Men sought not after knowledge where it was to be found, but sat gazing into the "smoke of perchance," dreaming that they might discern a form, or experimented with all manner of devices in search of a panacea. Dentistry fell like all other callings, and what little had been known was doomed and lost. Teeth were no longer considered in the light of organs to be rescued from destruction, but as amulets for warding off evil, or under varying circumstances as omens of good or bad.

Through medieval history the figure of darkness so frequently applied to affairs of those times, seems from the dental standpoint more appropriately to apply. Dental defects and deformities are mentioned in tones of pity, because they were considered without remedy. The familiar proverb which had its origin at that time "that a bad tooth was considered, of all things, the most desirable to be rid of," leads us into the secret of darkness. But the subsequent generations, with their speculations and projects, dispelled the magic smoke and the grand forms that were disclosed we will gladly speak of later in our study.

A good deal of the medical and surgical practice of this period was in the hands of religious orders, particularly of the Benedictines,* who made distinguished strides of advancement. In Paris a college was founded by the monks. The college was under the protection of St. Cosmas † and St. Damianus, two practitioners of medicine and surgery; the institution was known as the College de St. Come.† From the time Lanfranchi joined the school it attracted many pupils, and it maintained its independent existence for several centuries.‡ Early in the fourteenth century a council of the Roman church, held in Paris, decreed §"that monks and priests be forbidden to perform bloody operations," and surgery was again separated from medicine. By this division the barbers and bathers ‖ fell heir to the art and they continued to be the sole surgeons for several succeeding centuries. In those dark ages the barber's craft was dignified with the title of a profession,¶ being joined with the art of surgery. The French barber-surgeons were separated from the perruquiers and incorporated** as a distinct body in the reign of

*Encyclopaedia Britannica, vol. XXII., p. 675.
†Any Cyclopaedia.
‡See pages 93 and 94 for further notice on the clergy and medical science.
§American Cyclopaedia, vol. XV., p. 486.
See page 93 for Council of Tours.
‖Zell's Encyclopaedia, vol. I., p. 219.
¶People's Library of Information, p. 416.
Encyclopaedia Britannica, vol. III., p. 363.
**Ibid.

Louis XIV. In England barbers first received incorporation* from Edward IV. in 1461. During the reign of Henry VIII. the barbers were united with a company of surgeons, it being enacted "that the barbers should confine themselves to minor operations of blood-letting and tooth-drawing while the surgeons were prohibited from barbery or shaving."*

Some years since, the great American, Dr. Oliver Wendell Homes, in addressing a class of dental students, said:† "No longer ago than when President Holyoke, whose son, the venerable physician, some of us well remember, entered upon the duties of his office, and for years after that time, the London Company of Barber-Surgeons were holding their meetings at their hall in Monkwell street; and it was not till very near the middle of the last century that the surgeons were incorporated as a separate body. It was about the same time, that is, during the reign of George II., that the question was discussed in open court, before the chief justice of England, whether a surgeon was an 'inferior tradesman,' within the meaning of a certain statute of William and Mary. But we must remember in what contempt other of the most useful occupations were held so long as society

*Encyclopaedia Britannica, vol. III., p. 363.
American Cyclopaedia, vol. XI., p. 349.
†Missouri Dental Journal, vol. IV., p. 175.

was enslaved by its feudal traditions. Traffic and agriculture were scorned by the descendants of the Norman robbers, until they were starved into better views, and more civil language than they had inherited."

Early in the seventeenth century a jesting poet spoke of the barber-surgeon as:

"His pole with pewter basins hung,
  With rotten teeth in order strung,
  And cups that in the window stood
  Lin'd with red rags, to look like blood,
  Who shaved, drew teeth and bled a vein."*

About this same time there was a latent strife which eventually burst forth in an open contest between the surgeon and the barber. We read that the barber surgeons were separated from the mere perruquiers, and that the former were incorporated as a distinct body in the early part of the reign of Louis XIV. This success of the tonsorial trade gave them eminence in their own eyes, and being ambitious to continue to rise in public favor, made a desperate attempt to capture the dental art, and called themselves barber-chirurgeons;† but, alas, the few dentists who, at this period practiced the various departments of dentistry, saw danger in the barbers' new venture and so thoroughly protested the claim of the tonsors, on the ground "that the barbers practiced the science of dentistry

---

*Items of Interest, vol. XI., p. 556.
†Peoples' Library of Information, p. 416.

and were not regularly educated," that Louis XIV., in 1741, separated the barbers and dentists and made two distinct vocations.* The same grand change took place in England in 1785, under George II.†

The barbers on the one hand reluctantly yielded to the kingly proclamations, and have retained up to date a desire to perform the minor operations. Dr. Hunter says on this interesting struggle: "The memorial between the dental profession and the tonsorial art is still seen in the striped pole and basin sometimes seen projecting as a symbol in front of the barber shops."‡

The same was true with reference to the jewelers, but to a less earnest degree.

This separation between the humbler calling and the more dignified profession immediately gained for the science of dentistry a high social position, and has made the most marvelous progress known to any science in the annals of man.

The following extract from a popular medical journal portraying our art and its progress since the day when blacksmiths§ were its practitioners: "Certainly, a good dentist deserves to be called the friend of man. And therefore we read with pleasure that no branch of surgery has made so

---

*Peoples' Library of Information, p. 417.
†Encyclopaedia Britannica, vol. III., p. 363.
‡Encyclopaedic Dictionary, vol. II., p. 420.
§Items of Interest, vol. XI., p. 413.

much progress as dentistry has done; for, during many dark ages, with respect both to science and to practice, it was in a very backward state. Not very long ago, it is averred, blacksmiths were much in favor as operators in this department—a fact which seems to require explanation. It will perhaps be surmised that they were recommended for their work by their great bodily strength. But the obviousness of this account of the matter is delusive; the true theory must be sought in a more roundabout way. And if, in the first place, we remark that the blacksmith anciently discharged the functions of a 'farrier,' perhaps this will be thought not to cast much light upon the subject, but rather itself to need illumination. Remembering, however, that to the minds of our forefathers the offices of barber and surgeon seem naturally to go together, we cannot be surprised that to the same minds it should appear part of the fitness of things that the blacksmith who shod a horse should also doctor it. And, now, as Mr. Spencer would say, observe the implication. In doctoring a horse it must sometimes have been necessary to extract a tooth, and it was at once inferred that he who could extract a horse's tooth, a 'fortiori,' could draw a man's. And that he did often draw, to admiration, both the tooth and the man, may be imagined. Figure the blacksmith with his patient careering round and round the forge, emulating the dealings of—Achilles with Hector, and then listened to those who deride what they call the mere material civilization of the present day.   Great is the transition from the blacksmith's shop to the modern dentist's ingenious arm chair—we had almost written easy chair.   On the other hand, it may be that the need of dentists has much increased with civilization.   It is commonly believed that savages have excellent teeth; and although we are nowadays in the habit of suspecting such beliefs, this one seems probable, if we consider how necessary good teeth are to them.   To any one who is anxious to prove 'material civilization' a mistake, the inquiry may be suggested: What effect has the invention of knives and forks had upon the teeth of those nations that have condescended to adopt the use of them? For these pernicious utensils plainly render good teeth less a necessary of life than they were before, so that people with bad teeth now survive, transmit their degenerate natural weapons to their descendants, and so on."*

While the surgical portion of dental science was being looked after by the barber and blacksmith, the prosthetic branch was cared for by the skilled jewelerst of those times.   Jewelers in those early days were far ahead of the modern artists, since all the intricate gold and gem work, as also the

*Dental Cosmos, vol. XVI., p. 275.
†Preceding references.

engraving of same, was all done unaided by the
numerous machines and appliances at the com-
mand of the moderns. On this point Rev. Haines,
who has investigated antiquity, says: "Etrurian
and Roman jewelers were wonderfully skilled in
the rolling, smelting and manipulating of gold and
other precious metals in all their various forms.
In fact, Etrurian jewelry has been famous for
twenty-five hundred (2,500) years. I have seen
some specimens which were more than two thousand years old that would be difficult of reproduction to-day by any but the most skilled artificers."*
Hence Dental Prosthesis was in the safe keeping of
men who would to-day do credit to their own as
well as our profession.

But it was not until late in 1700 that the science
of Dental Prosthesis was eliminated from the jewelry
shop† and put in the hands of men who not only
understood the fundamental principle underlying
the science, but also thoroughly studied the human
mouth and its many adjacent connections. Oral
surgery on the other hand had not been completely restored to specialists or even medically-
skilled representatives; in every civilized country
of the earth the barber had been known to be defi-
nately connected with rude surgery, such as blood-

*Museum of Antiquity, p. 643.
†Items of Interest, vol. IX., p. 150.
Missouri Dental Journal, vol. IV., p. 175.

letting, dressing of wounds, and extracting of teeth
or the lancing of gums. In consequence of their
slight acquaintance with the rudiments of surgery,
the name barber-surgeon was usually applied to
those of the barber trade, who demonstrated some
surgical skill. In medieval Etruria the barbers
were usually men of liberal education, and hence
their intimate association with noted personages
gave rise to the Latin remark: "As inseparable as
musicus, tuturos et barba"—as inseparable as
the musician, teacher and barber. Thus for generations after prosthetic dentistry ceased to be
practiced by jewelers and watchmakers, oral surgery was firm in the grasp of the aspiring barber.

Years rolled upon years and generations after
generations passed beneath the sod, ere again the
dental specialist arose from among the down-trod-
den trades to rise to his former dignity.

The impulse which all departments of intellectual activity received from the revival in Europe of
Greek literature in the fourteenth century was felt
by medicine and its several branches, and their
corresponding practice was gradually improved.
The basis of medicine and surgery we learned was,
during the middle ages, dogmatic. "The medical
literature now brought to light," says Dr. Payne, in
his able treatise on medical progress, "including as
it did the more important works of Hippocrates

and Galen, many of them hitherto unknown, and
in addition the forgotton elements of Latin medicine and surgery, especially the work of Celsus,
was in itself far superior to the second-hand com-
pilations and incorrect versiohs which had formerly
been accepted as standards. The classical works
though still regarded with unreasoning reverence,
were found to have a germinitive and vivifying power
that carried the mind out of the region of dogma
and prepared the way for the scientific movement
which has been growing in strength up to our own
day. Two of the most important results of the
revival of learning were, indeed, the reawakening
of anatomy which to a large extent grew out of
the study of the works of Galen, and the investigation of medicinal plants, to which a fresh impulse
was given by the revival of Dioscorides and other
ancient naturalists. It was at first naturally imagined that the simple revival of classical, and especially Greek literature would at once produce the
same brilliant results in medicine as in literature
and philosophy. The movement of reform started
of necessity with scholars, rather than practicing
physicians and surgeons, more precisely with a
group of learned men, whom we may be permitted
for the sake of a name to call the medical humanists, equally enthusiastic in the cause of letters and
medicine. From both fields they hoped to expel
the evils which were summed up in a word—bar-

barism. Nearly all medieval literature was condemned under this same name; and for it the humanists proposed to substitute the originals of Hippocrates and Galen, thus leading back medicine to its fountain-head. Since a knowledge of Greek was still confined to a small body of scholars, and a still smaller proportion of physicians, the first task was to translate the Greek classics into Latin. To this work several learned physicians, chiefly Italian, applied themselves with ardor. Among the earliest were Nicholus Leonicenus (1428-1524); Giovanni de Monte (1498-1552); in northern Europe should be mentioned Gulielmus Copus (1471-1532) and Gunther, of Andernach (1487-1584). A little later Janus Hagenbut (1500-1558) and Leonard Fuchs (1501-1566), in Germany; John Kaye (1510-1572), in England, and Symphorien Champier (1472-1539), in France, carried on the work.

"The great Aldine press made an important contribution to the work by 'editions principles' of Hippocrates and Galen in the original. Thus was the campaign opened against medieval (middle age) medical writers, till finally Greek medicine assumed a predominant position."*

In the beginning of the sixteenth century medical, surgical and dental art and science were

revived, but to no marked degree. The anatomical research of Vesalius, later on others, and
prominent among them Fallopius Eustachius,
Pare, Hunter and Fox did much toward recreating
the medical art, which for upwards of ten centuries
lay dormant and unobserved.*

Indeed it is comparatively of late years that dentistry has occupied anything like a properly recognized position among the different departments of
medicine; for we learned that it was practiced to a
large extent as a super added means of livelihood by
persons engaged in some other pursuit, and without
any professional education or discipline whatever.
Under the very shadow of the famous European
universities, dentistry was professed by the blacksmith, barber, bather, jeweler, silversmith, monk
and even the cobbler. But matters were not destined to so remain, for the hospitable goddess of
liberty and enlightenment, whose natal day the
world shall ever cherish, unloosened fetters and
turned the page of progress. Educational matters
began to receive attention and the invention of
printing and the discovery of America, lead the
way to future triumphs.

---

*Any Complete Encyclopaedia.

ADDITIONAL BIBLIOGRAPHY.

Decline and Fall of the Roman Empire—Gib-
bon ; Revival of Learning—Hallam ; Lost Art—
Arnold ; Renaissance—Symonds ; Europe During
the Middle Ages—Hallam ; Period of Revival of
Learning—Payne ; History of the State of Rome
in the Middle 'Ages—Gregorovious ; Fall of the
Roman Republic—Merival ; Revival of Medicine—
Ægineta ; Revival of the Ancient Arts and Scien-
ces—Voigt ; Records of the Reformation—Pocock;
Ecclesiastical Annals—Raynaldus ; Studies in the
History of Renaissance—Symonds ; The Reforma-
tion—Dollinger ; History of the Papacy During
the Reformation — Creighton ; History of the
Roman Catholic Church—Dowling ; Works on
Printing and Progress of the Lost Ages—Harper.*

---

*See appendix for additional notes.

EUROPEAN DENTAL ART.

In the beginning of the seventeenth century dental science in Europe was revived, but to no marked degree; artificial dentures were beginning to receive attention and filling teeth with lead and putty opened the way to future conquests.

We will now briefly sketch dental progress among the French, Dutch, English, German and other people. In order to avoid repetition, we will treat in a general manner, the various methods, processes and materials used in the several European nations, under the cover of "European Dental Art;" and leave for further mention those inventions, discoveries and personalities as belong particularly to the individual country.

To sum up the general status of dentistry as practiced in Europe at the beginning of this century, we can safely trust in the able writer, Dr. F. Maury, who, in 1840, wrote as follows:* "We shall not here enter into a detail of the antiquity of Dental Prosthesis, but will merely say of it that the polished people of antiquity paid particular attention at all times to their teeth, and endeavored to repair their loss by mechanical means.

"Whatever may be the origin of odontotechny

---

*Dental Surgery—Maury, p. 179.

it is certain that this branch is carried to a greater degree of excellence at this present day (1840), than it ever attained in former times, particularly in our country (France), where this department of dentistry has become so good and so general as to be considered an art; the many advantages of which all classes of society have felt. It was Fauchard, who first in 1728, gave a treatise upon this subject (Prosthesis). Several French and foreign works have appeared since that time, and although incomplete, they furnish valuable information upon the subject now before us. They appear to us susceptible of important additions, and if we cannot flatter ourselves that we can furnish a perfect treatise upon this part of the science we can at least hope to point out the various improvements that have recently been introduced.

"The materials that have been used in constructing artificial teeth are the bones and the teeth of oxen, horse, sheep, stag and several other animals; ivory, mother-of-pearl, teeth of the hippopotamus or sea-horse, of the whale, human teeth, incorruptible teeth, made of mineral paste. Persons who have been deprived of their front teeth, have for a long time replaced them with artificial teeth made of white wax.   .

"We shall briefly examine these various substances, the number of which have been greatly reduced. We would however, here remark, that the teeth of the hippopotamus, the human, and the incorruptible teeth, are most generally used at the present time (1840).

"Bones of the Ox.—These bones, being entirely destitute of enamel, bear but little resemblance to the natural teeth, and are very porous; they become yellow, and decompose very soon. They have, however, for a long time been used for fabricating supports or bases resembling the gums. For this purpose the femur has been used, after having been cleansed in clay and exposed to the dew to whiten.

"Teeth of Oxen, Horses, Etc.—As we cannot give these the shape of the human teeth by means of the file, it is easy to detect them otherwise than by the absence of the enamel, which does not cover the surface of these teeth. Their color does not resemble that of the human teeth. If, however, we are obliged to use them from the want o human teeth, we should choose those of animals advanced in age, because of their central cavity being smaller than at a younger period of life; they are hence more solid and better adapted for the reception of pivots, by which they are to be attached to the artificial base.

"Ivory.—Sometimes parts of sets, and sometimes complete dentures are manufactured of this substance; but, like the preceding, it is not a good imitation of the natural organ. Ivory, being void of enamel, becomes yellow very soon in the mouth, and the saliva and mucous decompose it, after a time, in spite of the care that may be taken. In case we cannot procure a substance more resisting, as the tooth of the sea-horse, we should prefer the ivory of ·young elephants, and the central part of the tooth near the point. The grain at this point is more compact, and the ivory usually lasts longer than that of any other part of the tooth. Green ivory, that is, ivory taken from the teeth of animals that have been recently killed, is preferable.

"Mother-of-Pearl.—We need merely name this material. It is unfit to be used for this purpose, and has been seldom employed.

"Hippopotamus.—The use of the tooth of the hippopotamus in manufacturing teeth is of recent date, as ivory was, for a time, almost exclusively used ; but the inconveniences of the latter, already named, and the superiority of the sea-horse tooth, have induced dentists to abandon the use of ivory. At the present day the tooth of the hippopotamus is much used, both with and without the enamel.

"These teeth are obtained in commerce from Africa and the most distant parts of Asia. Such as are least hollowed out are considered best, as their ivory is more compact than those that are hollow. These teeth vary much in size, color, form and enamel.

"The incisors of the hippopotamus are short, semi-cylindrical anteriorly, contain a deep furrow, and are enveloped with enamel, the color of which, when polished, resembles that of the human teeth. Their semi-circular form enables us to carve from the same piece several teeth having enamel upon them. They sometimes contain deep furrows by which we are enabled to carve six or eight teeth thus shielded with this substance.

"The tusks of this animal are larger and longer than the incisors, and are curved like those of the wild boar. Their least weight is two pounds and a half. They sometimes weigh nine pounds, but this is uncommon. The teeth are flattened posteriorly, and convex anteriorly, and are covered with enamel only on the latter side. Their size enables us to form complete dentures of them, not enameled, or bases upon which enameled teeth are afterward to be attached.

"We should make use of such as have their internal substance compact, white and smooth. The best are white, round, enameled at their smallest part, and have not large ridges or deep depressions, and are not cracked in the direction of the curvature. To preserve them they should be kept in a humid place. When using them or working them, we should be careful not to expose them to the sun, fire or current air, as such exposure will tend to crack them, especially in such places as are not protected by the enamel

"If we cut a hippopotamus' tooth transversely

through its middle, we will perceive a furrow, whose depth depends entirely upon the age of the animal. We should as much as possible avoid this furrow in manufacturing pieces. If its use cannot be avoided, the piece should be so constructed that this defect may not be perceptible, as this part is yellow, and more easily acted upon by the secretions of the mouth.

"However perfect and beautiful may be the piece of this material used, its extreme whiteness, which at first pleases the eye, sooner or later is lost, and a bluish or yellow hue is assumed.

"Artificial bases are generally made of the tooth of the sea-horse, and human teeth are inserted into this; and when these bases are neatly carved and polished, they present a very good appearance.

"The incisors of the inferior jaw of the hippopotamus are called, improperly, in commerce, the teeth of the sea-cow. These teeth are round, and have no enamel, and when they are of a proper size, complete dentures are manufactured out of a single piece by its being cut lengthwise.

"Teeth of the Whale.—These teeth are sometimes in commerce mixed with those of the sea-horse. They are as strong as the latter, but differ very much from them in form, and in their durability. They may, however, be used in the manufacture of the bases when we cannot procure a substance more compact.

"Human Teeth.—Of the various articles used in replacing the lost organs, the human teeth, without doubt, merit the preference, since they are such as are given us by nature. We shall indicate the manner in which they should be chosen, and the various preparations they should undergo before insertion.

"These teeth are generally obtained from the mouths of persons who die in hospitals, and whose bodies are brought into the amphitheatre for dissection. The best are such as are not decayed or cracked, and have been taken from subjects between the ages of eighteen and forty years. The teeth at this time of life are firm and of the most desirable consistency, and are capable, for a long time, of resisting the destructive agents to which they are constantly exposed. The teeth of younger subjects are too tender, their canal is to large, and they are incapable of resisting deleterious influences. Those of old men are hard, but yellow and much worn, and crack very easily. We should prefer the teeth of adults which have been recently taken from the subjects. We should reject all such as are not entirely sound, or the cavities of which are red or black, as they very soon become black in the mouth and decay. It is true, that if a cavity be found upon the side of a tooth used for this purpose, we may drill the decay out, and insert a plug made of the tooth of the hippopotamus; but

these should not be used if sound ones can be procured.

"The teeth being chosen, we should preserve
them in such manner as to be able to use them at
any moment.   They should be extracted from the
subject with care, and portions of the alveolar process, periosteum, gum or tartar, that may adhere
to them, should be removed.   The extremities of
their roots should be pierced, and they should be
strung in the order in which they had been placed
in the mouth.   They should then be steeped for
seven or eight days in water, which should be
changed every twenty-four hours.   At the expira-
tion of this time, they should be again cleaned by
being rubbed with a piece of soft wood, as willow
or fir, wet and dipped in powdered pumice stone.
In this manner we can remove all foreign bodies
from the teeth.   If there should remain any stain
or spot, it should be removed with the file or grind-
stone.   They are next to be washed with soap and
water, and the process of cleaning is to be concluded by immersing them in alcohol.

"We generally use the eight superior teeth,
viz: four incisors, two "cuspidati" and two anterior
bicuspids.   It will be well, however, to procure the
eight corresponding inferior teeth, as they are
sometimes required.

"The teeth, being thus cleansed and assorted,
should be placed in a vase and covered with sand,

bran, fine grain, saw-dust, or anything capable of excluding the air, heat and cold.

"Some dentists preserve them in water or diluted alcohol. This is a bad practice, as they become yellow and crack afterward when exposed to the air. Others preserve them in equal parts of wax, chalk and oil; but this is inferior to the simple mode we have already recommended. When we cannot procure the eight teeth from the same individual, care should be taken to select such as harmonize well together. We should be cautious in using teeth that have been procured from cemeteries; for, after having remained in the earth for a time, their enamel is apt to be dull. Their bony substance, also, is likely to be yellow, or of a brown hue, which is the result of decomposition. While preparing teeth of this kind, they often break very easily, and when inserted soon become black and decay.

"When we wish to insert two or three of these teeth, we adjust them upon a platina or gold plate, if the alveolar border be not too much absorbed. If this be the case they should be mounted upon a sea-horse base, and secured to it by means of platina-rivets.

"Animal substances of which artificial teeth are made possess the inconvenience of being liable to a speedy softening and decomposition; and they tarnish and emit a disagreeable odor. We are

therefore obliged to renew them frequently. To
obviate this inconvenience, it has been proposed to
manufacture artificial teeth of earth, capable of being
hardened by means of heat, and enameled like porcelain. These teeth are called incorruptible."

Such, from the pen of an eminent French dentist, was the knowledge of artificial teeth early in
this century.

FRENCH DENTAL ART.

To the French scientists and the people generally is due much credit for having greatly encouraged dental art. The French system of dentistry
made rapid progress, and anon England, Germany
and other European nations copied nearly exclusively from the French artists. Ambroise Pare,
familiarly called the "barber-dentist," born 1517,
was an army surgeon, and he educated himself in
anatomical science, and in surgery was one of the
first great lights. He was surgeon successfully to
four kings of France, and was attached to the
French armies as surgeon-general as late as 1569.
"To Pare" says Sabine, "we owe the revival and
improvement in surgical practice." It was while
in the army in 1579, that he discovered the possibility of success in transplanting teeth. His success in the ventures proved him capable and he
turned his attention largely to this new surgical
discovery. He subsequently constructed artificial

dentures, having as bases gold and silver.   Pare exerted a great influence on surgical and dental arts. His extensive medical and surgical experience he published in 1562 and later, the editions having been translated into all modern languages.   In 1590, Pare died.

Following this eminent Frenchman, came others, all of whom contributed to the advancement of dentistry.   Thus Hemard, a French dentist, manufactured, in 1622, ivory dentures.   Petrie Torest invented the elevator in 1602.   Dupont, a Parisian dentist, in 1633, advertised himself as a specialist on "Implantation of teeth;" in 1735 the French Academy of Science announced the discovery of caoutchouc, which in various forms of preparation was used by dentists as a filling material.

In 1728 Dr. Fauchard, proposed as substitute for natural teeth, such as could be made of porcelain; this, however, was simply a key for some future inventor, as the proposed porcelain teeth were not yet in existence.

A curious old book has come to light, pertaining to primitive French dental art; that the book in question was popular in its time is demonstrated clearly as, in 1816, it had reached its fifth edition. It was entitled: "A Dissertation on Artificial Teeth" by M. De Chemant.   The chief point of interest in the book centers in a sheet of engravings illustrating the various types of dentures which

M. De Chemant, was prepared to supply to his
patients. He here clearly portrays a porcelain
bridge of ten teeth supported by four pivots, by
which it is fastened to the remaining roots of the
jaw. A simple tooth from this ingenious bridge,
would be a true representation of what we Americans call a Logan crown. M. De Chemant speaks
of these teeth as his invention and records the cir-
cumstance ; and to satisfy the curiosity of students
of primitive Dental Prosthesis will give his own
words on the subject:

"In 1788, when I exercised the profession of a
surgeon, I was consulted by a lady who had fallen
into such a state of weakness as produced considerable fears of her life. On approaching her I perceived a tainted odor which I thought proceeded
from her lungs, or her teeth which were black. I
examined her mouth and was struck with the bad
state of a set of human teeth implanted on the
base of a tooth of the hippopotamus. This set of
teeth removed, I perceived her mouth to be almost
entirely covered with small ulcers, and I had no
doubt that her disease was the effect of the putrid
exhalations which proceeded from the set of teeth,
and which corrupted the air she breathed ; what
confirmed this conjecture was, that after having
laid these teeth aside her health improved in a few
days. Perceiving that this lady would not do

PLATE II.

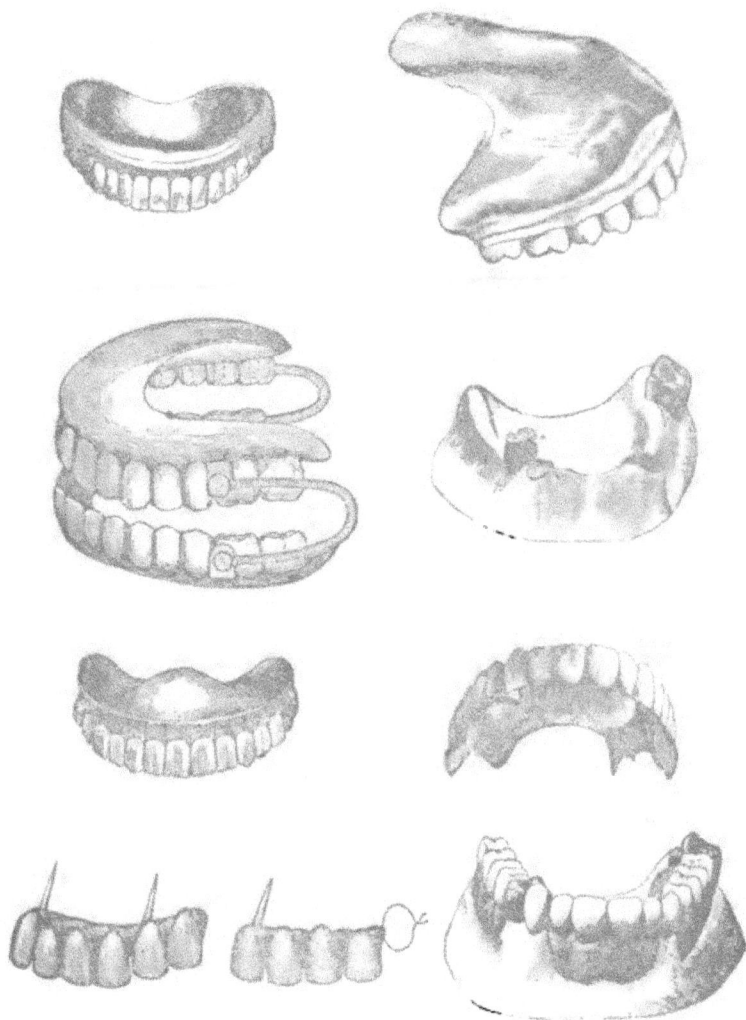

Specimens of Medieval Dental Art.

PLATE III.

without artificial teeth, I advised her to have
several sets of teeth at the same time, so that she
might change them often after having washed and
let them dry.   She did so, and her health became
re-established in the course of some months.   But
as the teeth of this kind required to be renewed
frequently they occasioned a great expense, and
notwithstanding their frequent renewal they
always produced a bad smell.   I was induced from
that time to reflect on the possibility and means
of making teeth and sets of teeth of durable and
incorruptible materials.   I examined almost all the
substances of the mineral kingdom, and at length
composed a paste which, when it was baked (porcelain) had every desirable advantage."

Now as a matter of fact, porcelain teeth were
invented by an apothecary of St. Germain, Ducha-
teau* by name.   He himself wore an artificial denture
of ivory and natural teeth, but found they rapidly
became tainted by the various disagreeable odors
emanating from his mouth, the porous animal substances becoming rapidly impregnated by the
effluvia.   This druggist called the attention of Mr.
Guerard to the discovery of a paste which, when
baked, became very hard.   The latter gentleman
undertook in 1776 to manufacture the substance
and with the aid of a dentist produced a porcelain
tooth.

---

*Items of Interest, vol. XIII., p. 13.

But to De Chemant is due the credit of perfecting the discovery or invention ; he bought the
right from the former and managed to attract the
attention of the French Academy of Science, which
at once appointed a sub-committee to examine the
teeth and their merits.   The committee reported
favorably, at the same time informing M. De Chemant where improvements were needed, and with
the aid of ·M. Dubois, dentist, the new substance
(porcelain) was considerably improved.

In 1805, Professor Lafargue published a book
on the practice and art of dentistry.   Dr. Debarre
is another early dental practitioner who deserves
much praise, since he early in the century published
a volume called "Prosthetic Dentistry." This is the
first book devoted especially to this great branch
of dental science.   This rare work was considered
superb and contained forty-two well executed
plates.   Subsequently to Dr. Debarre's volume,
many publications appeared in the French book
market.   .

In 1808, Fronzi,* a French dentist, constructed
a single enameled tooth to be eventually arranged
together in an entire set; he did great service in
improving enameled teeth.

It is claimed by modern dentists that Dr.
Fauchard, of France, in 1785, was the first doctor
to refer to gold-leaf as a filling material; and Dr.

---
†Dental Cosmos, vol. XXIII., p. 671.

Harris,* of America, is responsible for the statement that gold-leaf was first used for filling purposes in the early part of the eighteenth century. This on close investigation proves to be a mistake, as is clearly demonstrated in many pages of this work.

The following is a list of French authors, who had contributed to dental literature, prior to the year 1800.

Andree, 1784, 1790; Apius, 1751; Aurivillius, 1757; Auzebi, 1771; Bauhinus, 1660; Beaupreau, 1764; Botot, 1786, 1789; Bourdet, 1756, 1762, 1764; Brendel, 1697; Bunon, 1741, 1743, 1746; Courtois, 1775, 1778; Cransius, 1681; Cumme, 1716; Delabarre, 1800; Despre, 1720; Dubois De Chemant, 1789, 1790, 1796; Duchmin, 1759; Dupont, 1633; Fauchard, 1786; Finot, 1799; Fleurimon, 1682; Fronzi, 1798; Fouchon, 1775; Frank, 1692; Geraudly, 1737; Gilles, 1622; Grun, 1795; Hemard, 1582; Heslopp, 1700; Hilscher, 1748, 1776; Josse, 1800; Jourdan, 1761, 1766, 1756; Lecluse, 1750, 1753, 1782; Lemaire, 1784; Monier, 1783; Mouton, 1786; Ricci, 1790.*

These works were purely dental and this portrays how industrious and persevering the French scientists were relative to dentistry. The French have indeed done much toward establishing dental

---

*Harris' Principles and Practices of Dentistry.

literature, and the numerous dental journals published in France since 1857 go to verify this. The following are the names of the dental periodicals issued in the French language:*

L'Art Dentaire, founded 1857, Paris—Editors, Fowler and Pereterre.

Le Progress Dentaire, founded 1874. Paris—Editor, Stevens.

Le Cosmos Dentaire, founded 1876, Paris—Translated from American Dental Cosmos.

Gazette Odontologique, founded 1879, Paris—Editor not mentioned.

Annuarie Generale Desdenteste, founded 1880, Paris—Editor not mentioned.

L'Odontologie, founded 1880, Paris—Editor, Anbeau.

Revue Odontologique, founded 1880, Paris—Editors, Andrieu, Brasseur Damain, Gailland, Stevens, Colson and Quenot.

Revue Mensuelle des Maladies de la bouche—Editor, Saran.

Revue Odontolgia, founded 1882, Paris—Contumation of Gazette Odontologique, edited by Association Committee.

Le Monde Dentaire, founded since 1885, Paris—Editor, Rollin.

*These gentlemen have assisted me in getting a complete list of French journals: E. Kirk, J. Taft, H. J. McKellops, A. W. Harlan, L. Ottofy.

Gazette Odontologique de France, founded 1884, Paris—Editor, Quincrot.

Revue International d'Odontolgique, founded 1889, Paris—Editors, Dubois and Committee.

L'Avenir Dentarie, founded 1890, Paris—Editor, Delannay.

France is not noted for the number of her dental colleges, but for the high standard of these institutions. The two dental colleges which France has are both located in Paris and their names are: Institut Odontotechnique and Ecule Dentaire de Paris.*

DUTCH DENTAL ART.

The worldly matters of the fifteenth century were largely influenced by the Dutch people, who at this period of time were recognized among the leading spirits of the age. The Dutch territory, though small, exerted a powerful influence on the neighboring countries. Dutch vessels bathed in all the waters of the earth and early bore the honor of having the most formidable maritime power in the world. The people were no less active in educational matters, for we read that her schools of art attracted the attention of civilized Europe. In medicine particularly they stood as authority. These people were the first in modern history to introduce clinical instruction in hospitals,

---

*The author is indebted to Dr. A. W. Harlan for information relative to dental education in France.

and this departure was greatly developed, especially at Leyden in the hands of the celebrated Sylvius, who attracted students from all quarters of Europe. Soon the Dutch medical school became the most famous of the century. In dentistry, too, the already celebrated medical professors made advances, thus Andre Vesalius, a celebrated physician at Brussels in 1563, was the first to practice scarification of the gums, which he did on himself in order to facilitate the eruption of a wisdom tooth. In 1674 Nicholas Pulpius experimented largely with treatments for odontolgia and other dental affections. The famous naturalist, Lieuwenhock, in 1678 discovered the tubuli of dentine. And among those who strictly wrote on dental subjects prior to 1800 were: Brauer, 1692; Nicholi, 1799; Van Der Belen, 1782 ; Van Der Maessen, 1800; Van Veen, 1789; Valentine, 1727. The Dutch people from 1790 until within a generation ago suffered retrogression in educational matters, and the few strides of progress gained in the preceding centuries was forgotten and nearly entirely lost. A new era, however, has dawned, and education and learning is rapidly being revived.

Dental societies early in the last century were organized, the Societe Odontologique de Belgique being the senior ; dental journals were late in their appearance, but of late have been founded.

Revue Odontologique de Bruxells, founded 1884.

Revue Odontalogique de Belgique, founded 1885.

Dentistry at the present time receives marked attention, the prosthetic department taking the lead. Their schools of dentistry are institutes rather than colleges.

ENGLISH DENTAL ART.

The English people who were the prime factors in the medieval and early modern history, not only shaped the destinies of 'men, manners and nations,' but trades, occupations and professions also fell heir to their undaunted power and influence. Although dental art was revived in France it still required English talent to start the same on the highway of a profession. Prominent among the many medical students of the seventeenth century who devoted much attention to the revival of medicine and surgery are Sydenham Hunter, Fox and Blake. Thomas Sydenham was a graduate from Oxford and early demonstrated his genius. He entered the medical field and was known to be the most accomplished scholar in England on ancient medicines. As his model in medical methods, he repeatedly and pointedly referred to Hippocrates, and he has not unfairly been known as the "English Hippocrates." His influence on European medicine was great; his principles were welcomed as a return to nature by those who were weary of theoretical disputes. He was a strong advocate of specific medicine, and on this latter principle rests his fame. His knowledge of anatomy, physiology and therapeutics, inspired others to delve into these sciences, and thus indirectly Sydenham deserves much credit for having brought about the revival of "natural history of diseases." Private dissecting rooms and anatomical theatres were established, of which, perhaps, the most noted was Dr. William Hunter's school, London, inasmuch as it attracted students from all parts of the British Empire; it was in this school that the famous John Hunter received his first anatomical instructions. His brother William, with whom he was intimately associated, was one of the most brilliant exponents of medical science in England at that period. Under such eminent care, John made rapid progress, and before a score of years had rolled away he was the most famous physiologist and lecturer on surgery in the world. The dental profession of England particularly cherish his name, since in his work "Natural History of the Human Teeth," a treatise written and published in 1771, he laid the foundation of the English system of dental practice, from which the known world copied. Hunter, however, treated the subject anatomically and philosophically, rather than practically; the same may be said of nearly all the writers on this subject of that period. It would require many pages to merely enumerate the items of dental interest which are found in the works of Hunter, and it is doing him but partial justice to call him " Father of Modern Dental Surgery." Subsequent to the writings of Hunter came those of Blake, in 1798; and closely following in the footpath of the latter, came Dr. Fox of 1803.

These works, however, were not the first published in England on Dentistry ; the following are authors who, prior to 1800, furnished material for the English dental literature: Berdmore, 1770; Curtis, 1769; Herbert, 1778; Hurlok, 1742; Lewis, 1772; Ruspin, 1779; Timaeus, 1769; Tolver, 1752; Tuller, 1800; Walkey, 1793 and Woofendale, 1788.

We learned in the preceding chapter how and why dental art was revived and also the result of the struggle between the barbers and dentists. We are told that the earliest reference in modern times to the practitioner of dental surgery, as a dentist occurs in the Manchester (England) Times, in the later part of 1600. The following is the complete item in which the word dentist appears: "We are told that a clergyman who had taken temporary duty for a friend among us, and who had the ill luck to injure his false teeth during the week, the plate was sent to the dentist for repairs, a faithful assurance being given that it should be

duly returned by Sunday's post; but the dentist or
the post proved faithless. With the assistance
of the clerk the clergyman managed to stumble
through the prayers but felt it would be useless to
attempt to preach. He therefore instructed the
clerk to 'make some excuse for him and dismiss
the congregation.' But the feelings of the Reverend
may be better imagined than described when, in
seclusion of the vestry, he overheard the clerk in
impressive tones, thus deliver the 'excuse:' 'Par-
son's very sorry, but it is his misfortune to be
obliged to wear a set of artificial teeth. They
busted last Wednesday, and he ain't got them back
from London to-day as he was promised. I've
helped him all I could through the service, but I
can't do no more for him. It isn't any use him
coming out of there (pointing to vestry) and going
into the pulpit, for you wouldn't understand a word
he said, so he thinks you may as well go home."

Among the first men in England to be recognized as a dentist was Thomas Berdmore, dentist
to King George III.; he was an accomplished dentist, and in 1770 wrote and published his "Treatise
on the Disorders and Deformities of the Teeth and
Gums—Illustrated with Cases and Experiments."
Dr. Berdmore has the honor of being the first dentist to be employed as such in the royal family.
In his leisure he taught many medical students
who desired to practice dentistry as a specialty;

the famous John Woofendale having been one of his pupils.* Transplantation of teeth seemed to be a subject in which Dr. Berdmore was deeply interested and he is recorded as having been quite successful in this departure. He said on the matter that "The surgeons' art has taught that a tooth which has been partially or totally forced out of its socket may again be restored to its former situation and firmness, and may serve for use and ornament to the last period of life." And further "In the most favorable circumstances it is more than an equal chance that a tooth once extracted or beat out will fastened again."†

In a work published in 1774 by M. Patence, a dentist of London, there is the following on the same subject: "Of late it has been the practice to extract a tooth from the mouth of some poor persons and transplant into the socket of others for a few shillings."‡ Lefoule adds "It is selfishness of the rich who would deprive the poor of their teeth to replace their own."‡

The manner in which the operation was performed we gather from Dr. Hunters' work, "The Natural History of the Teeth:" "A fresh tooth when transplanted from one socket to another, becomes to all appearances a part of that body to

---

*Dental and Oral Science—Dexter, p. 5.
†Dental Cosmos, vol. XIX., p. 261.
Johnsons' Dental Miscellany, September, 1876.
‡Dental Cosmos, vol. XIX., p. 261.

which it is now attached, as much as it was to the one from which it was taken; while a tooth which has been extracted for some time, so as to lose the whole of its life, will never become firm and fixed; the sockets will also in this case acquire the disposition to fill up, which they do not do in case of the insertion of a fresh tooth. I would recommend to every dentist to have some dead teeth at hand, that he may have a chance to fit the socket. I have known these, sometimes to last for years, especially when well supported by neighboring teeth. But even this should not be attempted unless the socket is sound and pretty large, as the tooth can otherwise have but little hold."*

From various advertisements in English papers, we can get a fair idea of the general status of English dentistry in the latter part of 1700. The following of 1769 has: "M. Hamelton, Surgeon-Dentist and Operator for the teeth, from London, who cleans and beautifies the teeth and displaces all superfluous teeth and stumps, with the greatest ease and safety, and makes and sets artificial teeth from one single tooth to a whole set in so nice a manner that they cannot be distinguished from the natural." †

We can form some idea how awkward and unbecoming the dentures of these days were, when we

*Hunter, vol. I., p. 58, vol. II., p. 95.
†Dental Cosmos, vol, XVIII., p. 542.

read in Sheridan's, The Duenna, Act II., Sec., 3, these lines: "For her teeth,* where there is one of ivory, its neighbor is pure ebony, black and white alternately, just like the keys of the harpsichord."

The following from the pen of George C. Chance,† is of interest to those who desire a general idea of dental practice in the sevententh century: "In a recent visit to London I was permitted to copy from an almanac, now in the possession of James Parkinson, dentist, London, the following advertisement, which appeared in the Stationers' Hall Sheet, in 1709, in the reign of Queen Ann: 'Sets of teeth set in so well as to eat withal and be worn years together undiscovered. Also teeth cleaned and drawn by John Watts, operator; he applying himself 'wholly' to the business.' The John Watts of the above advertisement, was a partner of Thomas Berdmore, who held the appointment of dentist to the royal family, which appointment was subsequently held in succession by Mr. Parkinson's grandfather, his son and grandson, and then the appointment passed away. These all practiced their profession in the same house, in Regent Court. Mr. Parkinson being the last proprietor, disproved of the old house in 1858. The entries selected from his ledger of 1789 (in which

---

*Popular Quotations and Mythology—Carleton, p. 170.
†Dental Cosmos, vol. XVI., p. 107.

year the gross receipts of the business were
£2,886 or about $14,430), read:

' "One tooth leaded...........$ 2.50
One tooth gold-stopped....... 4.96
One new pivot tooth......... 4.96
Seventeen teeth drawn.
Gold plate for lower front teeth.
One impression taken.
Gold plate one tooth.
Two teeth stopped with gold.
Tooth drawn and replaced.
Four natural teeth put on plate.
Upper artifical set complete... 52.50 " '

The English dentists early turned their attention to periodical literature, though they may not
be able to boast of the number of their journals,
yet the standard of what few they have, is par excellence.   These are the professional journals of
Great Britain :*

Quarterly Journal of Dental Science, founded
1843, London—Editor, James Robinson.   Discon-
tined after 1886.

The British Journal of Dental Science, founded
1856, London—Editors, W. Blundell, Tomes,
Begg, Harrison, Cartwright, Saunders and Fox.
Still issued.

The Quarterly Journal of Dental Science (N. S.,)
founded, London, 1857—Editors, Rhymer, Per-

kins, Thompson and Robinson. Discontinued after 1859.

The Dental Review, founded, London, 1858—Editors, Rhymer and Hockley. Discontinued after 1864.

The Dental Review, New Series, founded 1864, London—Editors, Hockley & Rhymer. Discontinued after 1886.

Archives of Dentistry, founded, 1865, London—Editor E. Truman. Discontinued after 1886.

Journal of British Dental Association, founded, 1880—Editors' names not given. Still regularly issued.

The Dental Record, founded 1881, London—Editors' names not mentioned. Still published.

There are numerous English dental societies the most noteworthy of which is the well known Odontological Society of Great Britain, founded in 1867.

Dental education in England has, and is yet receiving great encouragement. In 1859 a College of Dentists was established, which was subsequently abandoned in favor of the L. D. S. of the Royal College of Surgeons. The nonpareil development of dental art and science in England is forcibly illustrated by the number, generally fair standing and good quality of their dental schools, devoted to educating young men for the profession. Although there are but few schools devoted exclusively to instruction in dentistry, those that deserve a special mention are: Royal College of Surgeons, dental department; King's College, dental department; Royal College of Surgeons of Edinburg, dental department; Owen's College Victoria University, dental department; National Dental Hospital and College; Guy's Hospital Dental School; Dental Hospital of London, and Dental Hospital of London Medical School; The Queen's College, dental department; Dublin Dental Hospital; Edinburg Dental Hospital and School; Exeter Dental Hospital; Glasgow Dental Hospital and School; Liverpool University, dental department; Plymouth Dental Hospital and School.

There are also dental infirmaries connected with the following hospitals: Charing Cross, London, Middlesex, St. Bartholomew, St. George, St. Mary, St. Thomas, University College and Westminster.

The difficulty in establishing dental colleges in England seems to be their dependence upon the universities on one hand, and the meddling of the government on the other. This at first thought may seem to be detrimental to the rapid progress and advancement of dentistry, yet by reflection we learn that a thorough preliminary education is required before entrance to the several universities, and thus the title of L. D. S. (Scentiate of Dental

Surgery) is conferred on intelligent men only, and this correct method of creating doctors must ever have an ennobling influence on the profession at large.*

GERMAN DENTAL ART.

The German people it seems, were reluctant about accepting the true worth of dentistry, either as an art or science. Their neighboring people had already acquired the knowledge necessary for the care of the dental organs, and foreign dentists made no impression on the stern teutonics. Germany was truly the home of the barber-surgeon, and even up to this late century the tonsors are looked up to as the proper dental surgeons. It is most surprising that a people like the Germans who exerted such a wonderful influence on all the arts and sciences, should so long remain dormant on a subject so highly interesting to humanity as dentistry. It has been said that the German students, scientists and philosophers devoted their energies in the study of all things foreign to man himself— and when we view the condition of the art of dentistry in 1700 we are lead to strongly believe in the charge made. Notwithstanding the fact that the Germans were late in appreciating the laws of dental preservation, lost time was seemingly regained through earnest study and German perseverance.

*We will speak of Canadian dental progress in connection with the advances in the United States of America.

Dental progress in German can be briefly told, since it is of recent development.

To a German dentist, Dr. Mesue by name, we should be grateful for re-inventing the process of filling teeth with gold-leaf. A German volume published in Frankfort in 1541, entitled: "Medicine for the Teeth, etc.," conclusively demonstrates this point as follows:

"Corrosio is a disease of the teeth when they get holes and hollows, happening mostly to the molar teeth, especially if they do not get cleaned after eating; for the victuals adhere, decay, produce bad, acrid fluids, that eat and itch into the teeth, and keep on doing so, until the teeth are entirely destroyed and one piece after another must, not without pain, drop off.

"This condition," according to Mesue, "is stopped and cured in three different ways : First, by purging ; secondly, by destroying the matter that hollows them out and eats them away ; this is done by boiling cockle, that grows in rye and wheat, with vinegar, and holding it in the mouth, or with vinegar in which capers-root with ginger is boiled. Thirdly, by getting rid of the hollow, which may be done in two ways: The first is to scratch and clean with a fine chisel, knife, file, or with any other instrument fit for it, the hollow and the parts attacked, and fill with gold-leaves, for the preservation of the remaining part of the tooth. The second is to use medicine, which is done by filling the teeth, after cleaning, with gall-nut and wild gallows-wood.   Or, take henbane-seed mixed with gum storax, and make with it a smoke through a funnel into the hollow tooth.   Or galbanum laid on hollow teeth mitigates the pains.   The pains are also quieted if the hollow teeth are filled with oppoponacum."

We learn that Frederick the Great employed the service of what he called a "Zahnarzt" (dentist) and this worthy gentleman's name was Pfaff.*   He was not educated in the purely prosthetic department and an appropriate name would have been "mechanical-dentist."   The only thing Herr Pfaff can be remembered for is the fact that he is the first dentist to describe a plaster-model of the mouth, and occurs among his private papers dated 1756.

Prior to the year 1800 there were published scores of works devoted to dental science, and among these the following: Naehere Pruefung der Etiologie der Zahnarbeit—Blumenthal, 1799; Einleitung zu den Wissenschaften eines Zahnarztes—Brunner, 1771; Zahnschmerzen—Glaubrecht, 1766; Praktische Darstellung aller Operationen der Zahnarzneikunst—Jeron, 1800; Sicherer Augen- und Zahnarzt—Krautermann, 1793; Belehrungen von der Wirkung electrischer Erschuetterung im

---

*Dental Cosmos, vol. XXIII., p. 671.

Zahnweh—Lentin, 1756; Abhandlung von den gewoehnlichen Zahnkrankheiten—Meyer, 1778; Zahnfleisches und der Kiefer Krankheiten und Heilart—Pasch, 1767; und Behandlung der Zaehne—Mesue, 1541.*

When the German people did wake up to the usefulness of the dental organs they seemed over anxious to demonstrate to the world that although last to acknowledge dentistry as a healing art they like the Holy book tells "the last shall come first." Dental journals immediately sprung into existence and did eminent service to the dental profession in Europe. A list of their periodicals reads:

Der Zahnarzt, founded 1855, Leipzig—Editor, Schmedicke. Discontinued in 1856.

Deutsche Vierteljahrsschrift, founded in 1861—Editors, Nedden and Heider. Discontinued in 1862 and published as:

Deutsche Monatsschrift fuer Zahnheilkunde, founded in 1862, Leipzig—Editors, Baume and Stolper.

Correspondenz-Blatt fuer Zahnaerzte, founded in 1871—Editors' names not given. Still published.

Zahnaertzlicher Almanach, founded in 1876—Editor, A. Peterman.

Notizen-Kalender, fuer Zahnaerzte, founded in 1877—Editor, Safford. Discontinued in 1880.

---

*Maury's Dental Art, p. 267.

Der Zahnaerztliche Bote, founded in 1879—Editor's name not given. Discontinued after 1886.

Die Zahntechnische Reform, founded in 1881—Editor, Pawelz. Still issued.

Vierteljahrsschrift des Vereins Deutcher Zahnkuenstler, founded in 1881—Editor, Jantsch and Polscher. Discontinued in 1883 and published as:

Deutsche Monatsschrift fuer Zahnheilkunde, founded in 1883—Editor, Parreidt.

Centralblatt fuer Zahnheilkunde, founded in 1883—Editor, Goldstein. Discontinued in 1886.

Zahnaerztliches Wochenblatt, weekly, founded—Editor Andreae. Still published.

Journal fuer Zahnheilkunde, weekly, founded in 1889—Editor, Richter.

Relative to their schools the eminent dentist, Prof. Miller, of Berlin, says:* "In 1869 a law was enacted in Germany which in effect entitled any one to practice medicine or any specialty of medicine without any qualification whatever. While this law did not affect the practice of medicine, it had a marked effect upon the practice of dentistry. A class of 'teeth-artists' appeared who entered upon the practice of dentistry from the standpoint of the mechanic. At the present time, while there are not more than one thousand qualified practitioners, there are four or five times that many who have no qualifications except such chance informa-

*Dental Cosmos, vol. XXXII., p. 992.

tion as had been picked up. Some of these men become skillful in mechanical directions and compete successfully with men who have graduated from dental colleges. It is only within recent years that dental students received instruction in practical dentistry. Formerly they were launched upon the world utterly unprepared, and therefore the 'teeth-mechanics' were able to compete with the graduates. This has had the effect of stimulating the dentists to better qualify themselves for intelligent practice. Of course many German dentists made for themselves enviable positions in practical as well as in scientific dentistry, although the mass of the profession, as already set forth, was in practical dentistry far below the standard reached in America. Four or five years ago a few American dentists located in Germany and have exerted a marked influence for good. Within the last six years schools of dentistry have been established as departments of several German universities, and in the school, in which I am connected, an earnest endeavor has been made to emulate the thoroughness of instruction in American schools." After alluding to the accomplishments in bacteriology by Koch, others and himself he concluded by saying: "The physicians in Germany are taking greater interest in dental matters, but a higher standard of education is needed to develop a more intelligent appreciation of the importance of such facts as have been alluded to.   At present there is virtually a three years' course instead of a two years' course which prevailed a few years ago. Physicians as well as dentists have recognized the fact that medicine meets upon common grounds with dentistry, and that it would be as impossible to separate medicine from dentistry as to separate the mouth from the alimentary tract."

The other European nations modeled after the French and English dental art with the exception of Germany, which land of learned physicians and surgeons made additional progress in dental science. Perhaps the only item pertaining to the German art of Dental Prosthesis is the mysterious discovery near Bologna, Germany.* Some years since while some workmen were at work digging a ditch through an old God's acre, were obliged to excavate many graves.   The dead of course had differentiated to earth, but dozens of artificial dentures still remained and were in good repair.   The dentures were made of ivory and bone, while the plates were of solid gold.   The superintendent of the excavations, thinking little of the find, sold the old dentures to a goldsmith, who in turn thought slightly of his purchase and remelted and refined the precious metal.   Whether these gold dentures were the workmanship of ancient dentists or the production of medieval jewelers, is beyond the power of this generation to prove.

AUSTRIAN DENTAL ART.

The first dental school on the continent of Europe was opened in Vienna on the 14th day of December, 1874. It bore the modest name of "First Dental School of Vienna."* The course was divided into a preliminary one of four weeks and a regular course of six months. "Students and doctors of medicine will be admitted," say their catalogue, "to the lectures, and doctors only to the labratory. The fee is 300 flourins. The school will not confer degrees, the government gives the title 'Magistrate of Dentistry.'"

In 1885 a dental journal was founded called, Oesterreichisch-Ungarishe Vierteljahrsschrift fuer Zahnheilkunde—Editors, Schmid and Weiss. It is still published. Aside from a few dental organizations of Austria, the dental progress is similar to the countries already described. The following countries have one or more dental schools and journals:

SWITZERLAND.—Schweizerische Vierteljahrsschrift fuer Zahnheilkunde, Zuerich—Editors, Redard and Frick.

ITALY.—L'Odontologia, Palermo—Editor, Riballa Nicodemi. Giornal di Correspondenza pei

---

*Dental Cosmos, vol. XVII., p. 222.

Dentisti, Milan—Editors, Dott and Coulliaux. La Scienza Dentaria Revista, Florence—Editor, Cianchi. La Progress Dentistico, Milan—Editor, Platschick. La Reforma Dentistica, Naples—Editor, Cali.

SPAIN. — Le Odentologia, Cadiz — Editor, Aguilar.

RUSSIA.—Messager Odontologique, St. Petersburg—Editor Sinitzin (A. P.) Suboviachebny Vestnick, St. Petersburg—Editor, Sinitzin (H).

HUNGARY.—Odontoskop, Budapest — Editor, Iszlai.

SCANDINAVIA. — Skandinavisk Tandlægeforenings Tidskrift, Copenhagen—Editor, Christensen. Den Norske Tandlægeforenings Tidende, Christiana—Editor, Seel. Nordisk Kvartalskrift for Tandlægekunst — Editor's name not given. Tidskrift for Tandlæger—Editor, Carstens.

AMERICAN DENTAL ART

Dentistry had as yet received no great impetus, nor had the art met the people who were to restore it to its ancient professional dignity, and later in their acquaintance introduce the mere art to its superior companion the science. But latent beams of the science were destined to appear and we can be proud to say that the initial steps toward progress and attainments were taken in our own dear native land. It was during the period covered by our war for independence, something over one hundred years ago that dentistry was introduced into America; but it was yet in its crude state and absent of mature modern development.

Believing it appropriate to make some particular allusion to those dentists who were the pioneers of the profession in this country, and who laid, solidly and durably, the foundations of the present superstructure of dentistry, we shall briefly treat of these patriarchs.

In October of the year of 1766, there arrived in the United States, from England, Mr. John Woofendale.* This gentleman was a regularly educated dentist, having been instructed by Dr. Thomas Berdmore, whom we learned was the dentist to King George III. He is the first dentist, so

---

*Dental and Oral Science—Dexter, p. 7.

called and practicing as such, of whom any record
can be found, as having visited this country. Mr.
Woofendale commenced practice in New York soon
after he arrived. He also practiced for a short
period in the city of Philadelphia. While in this
country he did dental work for many prominent
Colonial-Americans, among them Mr. William
Walton of New York, for whom he constructed an
entire double set of artificial teeth, which is believed
to be the first full set inserted in America. But
the doctor did not long remain in the colonies;
either because he did not receive sufficient practical
encouragement, or from some other unexplainable
cause he returned in March, 1768, to England his
native home. He divided his time between New
York and Philadelphia for a period of eighteen
months. In 1785 he returned from London to
America and having purchased a farm in New Jer-
sey, retired, and in 1828, at the age of 87 years,
died.

From the time when John Woofendale left for
England until late in the colonial confederacy
there was not, as far as can be determined, a regularly practicing dentist in this country. Such, in
brief, was the general status of dentistry at the
birth-time of our grand republic. "When we look
back to that condition of the science," says Dr.
Dexter,* "we may and do experience, a just feeling

* Dental and Oral Science, Dexter, p. 6.

of pride and perceive matter for gratulation in the
giant strides of improvement made manifest by the
slightest comparison of ' then ' and ' now.'   Such a
retrospect is the best and surest means of encouragement to continue in the path by which we have so
rapidly and agreeably advanced, and will insure, it
is certain, a further and equally great elevation of
our profession among the liberal sciences in the
future."

In the month of July, 1778, perhaps the darkest
period of the American revolution, the French fleet
with reinforcement reached our shore. Among
the proud French soldiers was one named Dr.
Joseph Lemaire,* who came expressly to this country to battle for American independence.   This
patriotic Frenchman, Lamaire, soon became an
intimate associate of both Washington and Lafay-
ette, and while fighting with them for the independence of the colonies he often relieved the
suffering soldiers of the pangs of odontalgia.   Not
only did he exercise his skill to secure for the
revolutionary veterans freedom from physical suffering, but he too seized his sword and shouldered
his gun and played an active part in that memora-
able strife for human liberty. While the French
and American armies, in 1781–2, were in winter

---

*Annals of Philadelphia—Watson, vol. I., p. 179.
Dental and Oral Science, Dexter, p. 7.
Dental Advertiser, vol. II., p. 2.
Western Dental Journal, vol. I., p. 995

quarters, side by side, near Providence, Rhode
Island, Dr. Joseph Lemaire, by permisssion of
Count Rochambeau, the commanding general,
taught the dental art to Josiah Flagg, then eighteen years of age, and James Gardett aged twenty-
five. Lemair's practice was not limited to the
soldiers only, but he did dental operations for the
people in the immediate vicinity.

Dr. Hayden speaks of him as follows :* "The
first hints that were afforded or opportunities
offered to any person to obtain a knowledge of the
profession were, we believe, through a French dentist, by the name of Lamaire, who offered his
services to the public during the revolutionary war.
* * * He was not without some pretentions to
skill in practical operations, especially in transplanting teeth. * * * He likewise undertook
to instruct some two or three persons in the profession, which may be considered as the origin or
commencement of dentistry in this country." In
the winter of 1785-6, Lemaire transplanted one
hundred and seventy teeth, and not one succeeded;
however this practice of his fell into disfavor,
whether from lack of skill on the part of the Frenchman or from the very wide prevalence of infectious
diseases liable to be transplanted, we cannot say.
Watson in his "Annals of Philadelphia," say, that
" Dr. Lamaire arrived in this city in 1784, and

*Dental and Oral Science, Dexter, p. 8.

here continued the practice of dentistry. From all accounts of the Frenchman, we are led to believe that he was a polished gentleman and had received an excellent schooling in France.* He certainly must be remembered as a painstaking and studious professional, since in the year 1812 he published his first dental work, entitled "Le dentiste des Dames," Paris, 1812. This work was published in four editions, namely, 1812, 1818, 1824 and 1833. He issued "Deux Observations d'anatomie pathologique sur les dents," Paris, 1816; he wrote a volume known as "Histoire naturelle des maladies des dents de l'espece humaine" (translated from the English work of Joseph Fox, Paris, 1821); following this he contributed to the early dental literature a book in three volumes entitled "Traite sur les dents physiologie, Pathologie," Paris, 1822 and 1824.

About the same time Dr. Isaac Greenwood† emigrated from Great Britain and settled in Boston, where he did an extensive business until his death. How or of whom he learned the dental art does not appear on record. ' He had two sons, Clark and John, who learned of him, and of these two dentists we will read later.

Shortly after the establishment of the United States as an independent nation, an Englishman,

Dr. Whitelock* by name, emigrated to our shores and practiced dentistry in the New England states. How long he continued is not known, but positive evidence can be obtained that he landed in 1784. Of him Dr. Hayden says: ''Dr. Whitelock is a gentleman of polite address and accomplished manners, who, about the same time as Lemair, or shortly after, entered this country as one of a company of theatrical performers who were induced to come here through a rage for theatrical performances."

Josiah Flagg,† we learned in the preceding pages, obtained his knowledge of dentistry from Joseph Lemaire. Both Lemaire and Flagg were soldiers in the American cause, the one in the French army and the other holding a major's commission in the American. Thus, while in winter quarters in 1781, in his leisure time, Flagg was under the immediate instruction of the French surgeon-dentist. Upon the close of the war Dr. Jossiah Flagg settled in Boston, where he practiced dentistry. When the second war opened with England in 1812 he again became a strong advocate for the Union cause and enlisted in the army. But early in this struggle Dr. Flagg was taken prisoner and taken to England, where, on parole,

---

*Encyclopaedia Britannica, Am. Sup., vol. III., p. 78.
 Dental and Oral Science, Dexter, p. 8.
†Ibid, p. 9.

he made the acquaintance of Sir Ashley Cooper and
assisted him in surgical operations at Guy's Hospital, London.   After staying in England several
years he returned to America to resume his practice in Boston.   An extract from a circular which
Dr. Flagg distributed in the city of Boston and its
environs is interesting since it portrays fairly the
dental practice of 1785 and later.

We have before us this interesting document
which gives quite acurately the degree of proficiency which had been reached in dentistry towards the close of the last century.   It consists of
an advertisement issued by Dr. Josiah Flagg, surgeon-dentist, who informs the public "that he
practices in all the branches with improvements
[i. e.], Transplants both live and dead teeth with
great conveniency, and gives less pain than heretofore practiced in Europe or America. * * *
Sews up hare lips, * * * cures ulcers. *
* * Extracts teeth and stumps, or roots with
ease. * * * Reinstates teeth and gums that
are much depreciated by nature, carelessness,
acids, or corroding medicine. * * * Fastens
those teeth that are loose (unless wasted at the
roots), regulates teeth from their first cutting to
prevent fevers and pain in children, assists nature
in the extension of the jaws, for the beautiful
arrangement of the second set, and preserves them
in their natural whiteness entirely free from all

scorbutic complaints, and when thus put in order and his directions followed (which are simple) he engages that the further care of a dentist will be wholly unnecessary. * * * Eases pain in teeth without drawing. * * * Stops bleeding in the jaws, gums, or arteries. * * * Lines and plumbs teeth with virgin gold, foils or leads. * * * Fixes gold roofs and palates and artificial teeth of any quality without injury to any independent or natural ones, greatly assisting the pronunciation and the swallow when injured by natural or other defects. A room for the practice with every accommodation at his house, where may be had dentifrices, tinctures, teeth and gum brushes, mastics, etc., warranted approved and adapted to the various ages and circumstances ; * * * also chew-sticks, particularly useful in cleansing the fore teeth and preserving a natural and beautiful whiteness ; which medicine and chew-sticks are to be sold wholesale and retail that they may be more extensively useful.

"* * * Dr. Josiah Flagg has a method to furnish those ladies and gentlemen or children with artificial teeth, gold gums, roofs, or palates that are at a distance and cannot attend him personally.

"Cash given for handsome and healthy live teeth at No. 47 Newburg Street, Boston (1796)."*

*Dental Cosmos, vol. XVII., p. 669.

The document is ornamented in one corner by
formidable and antiquated instruments, while in
the other are to be seen tooth-brushes quite of the
modern pattern.   It has been preserved by a de-
scendent of one who, as may be seen on the back,
purchased a brush and tincture from Josiah Flagg
in the year 1800.

In 1785 Dr. Flagg issued an advertisement
which also portrays a few hidden items as to early
American dental practice.   It reads: "Dr. Flagg
transplants teeth, cures ulcers and eases them from
pain without drawing; fastens those that are loose;
mends teeth with foil or gold to be as lasting and
useful as the sound teeth, and without pain in the
operation; makes artificial teeth and secures them in
an independent, lasting and serviceable manner.
Sews up hair-lips, and fixes gold roofs and palates,
greatly assisting the pronunciation and the swol-
low.   * * *   Cuts the defects from the teeth and
restores them to whiteness and soundness, without
saws, files, acids and such abusives as have shame-
fully crept into the profession, and which have destroyed the confidence of the public.   Sells, by
wholesale and retail, dentifrices, tinctures, chew-
sticks, mastics, teeth and gum brushes, suitable
for every age, complaint and climate, with directions for their use."

Dr. Josiah Flagg died in Boston at the age
of fifty-two.

James Gardette * was born in France in 1756. He studied medicine for two years in Paris (1773–1775), and immediately afterward spent two years in the hospital practice at Foulon; he was commissioned as surgeon in the French navy in 1777. He was instructed in dental art by M. Taudinier, a dentist of high standing in Paris. He came to America in 1778, landing at Plymouth, Mass. He subsequently resigned his commission in the French navy and adopted this country as his home. When the French fleet and army under Count Rochambeau arrived at Newport, 1781, Gardette visited that town and made the acquaintance of Lemaire, of whom he received further instructions in dentistry. Both he and Lemaire found occasion to do dental services for many revolutionary generals and their subordinates. In the autumn of 1783, he went to New York where his professional success appears to have been but slight. In 1774 he removed to Philadelphia, where he continued in a very successful dental practice until 1829, when he returned to France.

"Dr. Gardette's name," says Chapin A. Harris, "will always be prominent among those of the best American detists. As an operator Mr. Gardette displayed great judgment, care and dexterity." He was the first to substitute flat clasps for liga-

*Dental and Oral Science—Dexter, p. 9.
Dental Advertiser, vol. II., p. 2.

tures or wires in artificial dentures.   He invented the " mortise plate " to which the teeth are secured by means of gold pins, and which permits the tooth to rest upon the gums instead of the gold plate. The first application of the principle of suction or atmospheric pressure has been attributed to him. He was also an earnest advocate of the practice of substituting gold-foil for lead in filling teeth and related that he had at one period prepared gold-foil for his own use from Dutch ducats, when no sufficiently skilled gold-beater could be had.   So far as can be learned he published but one work, titled "Transplantation of Human Teeth."   In 1829 he sailed home to France and in August, 1831, at the old age of seventy-five, he died.

CLARK AND JOHN GREENWOOD.

These two gentlemen were the sons of Dr. Isaac Greenwood, and both learned their father's chosen profession.   Little indeed is known of Clark; it is claimed he was born in England and when a lad accompanied his father to America.   After having acquired a fair knowledge of dentistry he left Boston and journeyed to New York City where, in 1778, he opened an office and continued the practice.

John Greenwood was born in Boston and from all accounts was the first native born dentist.

Young Greenwood, at the early age of fifteen

enlisted in the American army and fought in the
battles of Bunker Hill and Trenton, and was also
in the expedition to Canada under Gen. Arnold.
He afterwards entered the naval privateer service,
in which he remained until the close of the revolutionary war; when finding himself out of employment, he applied to his brother Clark, who was in
New York City practicing dentistry, but here we
learn he received no encouragement. He then
embarked in the business of nautical and mathematical instrument making. Soon after having engaged
in this business Dr. Gamage of New York requested
young Greenwood to extract a tooth for one of his
patients, which he did very successfully. This was
the commencement of his practice as a dentist.
He, however, continued at the manufacture of
instruments and added ivory turning to the trade ;
he practiced dentistry as opportunities afforded,
the demand for his services, however, in this last
departure soon increased to such an extent that he
was compelled to abandon his other occupations.
His practice grew rapidly and was obliged to procure assistance, which he did; Dr. William Pitt
and his brother Clark acted in the capacity of professional assistants. Dr. John Greenwood was
well informed in the surgical department of his
profession, as an example of which is quoted the
fact ''that during his practice in the treatment of
a diseased maxillary sinus, he perforated this

cavity from the socket of an upper molar and
effected a cure."* It has been said that he was the
first in the United States to strike up gold-plates
to serve as bases for artificial dentures, without a
knowledge of it having been done across the sea.
During his professional career while in New York
he immortalized himself and profession as well by
carving from the tusk of the hippopotamus a full
set of artificial teeth for the great American "who
was first in war, first in peace, and first in the
hearts of his countrymen." This set of teeth was
secured by spiral-springs and was intricate in
its appearance as compared with our later dentures.   The denture in fact caricatured the lower
portion of his noble face, and they gave him much
discomfort, and as Senator William Maclay of that
time said,† "his voice is hollow and indistinct,
owing, I believe, to artificial teeth."

The material used in the set was hippopotamus
ivory. The lower plate was made of one solid
piece, teeth and base being carved together; the
upper denture required greater skill, and was made
with the plate separate and the teeth riveted
to it with fine gold rivets.   The General had several sets of teeth, but the only one that gave him
any comfort were those made by Greenwood.   A
Swiss artist of New York City also produced den-

*Dental and Oral Science—Dexter, p. 11.
†Items of Interest, vol. XIII., p. 183.

tal substitutes for Washington, as the following
recently written regarding this mechanism obviously tells: "The plate or framework which held
the teeth in his mouth was made of iron, and after
Washington's death were sent to the New York
Loan Exhibition in aid of Washington's memorial
arch, but was deemed by the committee too horrible to display; so they locked it up in a safe. No
one could have dreamed what it was unless it had
been labeled; most spectators believed it to be a
colonial rat trap."*

Probably had it not been for the ingenuity of
the first American dentist, Dr. Greenwood, the
basic structure of this glorious country would have
lacked completeness by having at an early date
lost its most devoted father. A block of marble
might with propriety have been cemented into the
monument reared in memory of Washington, and
in it carved in glittering letters the name of DR.
JOHN GREENWOOD.

The following letter of Dr. John Greenwood to
Gen. George Washington, as an historical treasure
is fully worthy of space. It reads:†

"NEW YORK, Dec. 28, 1798.

"SIR: I send you enclosed two setts of teeth,
one fixed on the old barrs in part, and the sett you
sent me from Philadelphia, which, when I received

*Items of Interest, vol. XIII., p. 183.
†Magazine of American History, vol. XVI., p. 294.

was very black, occasioned either by your soaking
them in port wine, or by your drinking it. Port
wine being sour takes off all polish and all acid
has a tendency to soften every kind of teeth and
bone. Acid is used in coloring every kind of
ivory, therefore it is very pernicious to the teeth.
I advise you to either take them out after dinner
and put them in clean water and put in another
sett, or clean them with a brush and some chalk
scraped fine. It will absorb the acids which collect from the mouth and preserve them longer—I
have found another and better way of using the
sealing-wax when holes are eaten in the teeth by
acid, etc. First observe and dry the teeth, then
take a piece of wax and cut it into small pieces as
you think will fill up the whole; then take a large
nail or any other piece of iron and heat it hot into
the fire, then put your piece of wax into the hole
and melt it by means of introducing the point of
the nail to it. I have tried it and found it to con-
solidate, and do better than the other way, and if
done proper it will resist the saliva. It will be
handier for you to take hold of the nail with small
plyers than with tongs thus, the wax must be
very small, not bigger than this (*). If your
teeth grow black take some chalk and a pine
or cedar stick, it will rub off. If you want your
teeth more yellow, soak them in broth or pot
liquor, but not in tea or acids. Porter is a good

thing to color them, and will not hurt but preserve
them, but it must not be in the least pricked—you
will find I have altered the upper teeth you sent
me from Philadelphia. Leaving the enamel on
the teeth don't preserve them any longer than if it
was off, it only holds the color better, but to preserve them they must be very often changed and
cleaned, for whatever attacks them must be re-
pelled as often, or it will gain ground and destroy
the works. The two setts I repaired is done on a
different plan than when they are done when made
entirely new, for the teeth are screwed on the
barrs, instead of having the barrs cast red hot on
them, which is the reason, I believe, they destroy
or dissolve so soon near the barrs.

"Sir, after hoping you will not be obliged to
be troubled very soon in the same way,

"I subscribe myself,

"Your very humble servant,

"JOHN GREENWOOD.

"Sir, the additional chare is fifteen dollars.

"P. S.—I expect next spring to move my
family into Conneticut State. If I do I will write
and let you know, and whether I give up my present business or not I will, as long as I live, do anything in this way for you if you require it."

The following is Washington's reply:*

---

*Magazine of American History, vol. XVII., p. 438.
Dental and Oral Science—Dexter, p. 11.

MOUNT VERNON, 6th Jan., 1799.

"SIR: Your letter of the 28th ult., with the parcel that accompanied it, came safe to hand, and I feel obliged for your attention to my requests, and for the directions you have given me.

"Enclosed you have bank-notes for fifteen dollars, which I shall be glad to hear has got safe to your hands. If you should remove to Connecticut, I should be glad to be advised of it; and to what place, as shall always prefer your services to that of any other, in the line of your present profession. —I am, sir,

Your Very H'ble Servant,

"GO. WASHINGTON."

He carved one in 1790 and another set in 1795. In a letter which has been preserved, "The Father of his Country" complains, in a dignified way, that his teeth hurt him and do not very satisfactorily serve the purpose for which they were designed. He complains of their "bulging out the upper lip," and "causing the jaw line to protrude, giving the face an unnatural appearance."[*]

Thus Dr. Greenwood is best known to the profession of to-day through the fact of his having been the dentist of the first President of the United States. As near as can be learned Dr. John Greenwood died in New York City in 1816.

---

[*]Demorest Family Magazine, December, 1891.

Horace H. Hayden* was born Oct. 13, 1768, in Winsor, Conn. At the age of fourteen he went to sea as a cabin-boy, voyaging to the West Indies. In 1784 he abandoned sailing, and being thrown on his own resources by the poverty of his parents, he became apprenticed to an architect, which business he followed until his twenty-fourth year, when, being in New York, and having occasion for the professional service of a dentist, he visited the office of Dr. Greenwood. While under treatment he determined to study dentistry. He soon procured the few dental books then in existence, and not apprehending any deficiency in the mechanics he directed his undevided study to the calling. He settled in Baltimore in 1804 with little practical knowledge of the art and science. Dr. Hayden's previous education was hardly calculated to further his professional career, but being a man of considerable energy and ability, he, by dint of hard study, soon mastered his text-books, applying himself sincerely to anatomy, physiology, surgery and general medicine. His proficiency in these studies soon attracted the attention of the medical profession, both locally and generally, and secured him a recognition in the latter profession; having merited the honorary degree of "Doctor of Medicine" by both the University of Maryland and the

*Dental Advertiser, vol. II., p. 3.
Dental and Oral Science—Dexter, p. 13.

Jefferson College of Philadelphia. In 1825 he was invited by the former institution to read a course of lectures on dentistry before its medical class; he accepted the invitation and thus inaugurated the first move in the direction of oral teaching in dentistry in a college. He was one of the founders of the Baltimore College of Dental Surgery and also of the American Society of Surgeon-Dentists; and one of the editors of the American Journal of Dental Science, the first dental periodical ever issued. Besides having written many articles on dentistry and medicine, he contributed valuable theses on geology.

He was surely a man of unusual strength of character and few pioneers in dentistry did more to shape the future profession. He died Jan. 26, 1844, at the age of seventy-five.

Edward Hudson,* who was born in Ireland in 1772, had been thoroughly educated by his cousin and adopted father, a dentist of the highest position in Dublin. At the age of thirty-three he emigrated to America and settled in Philadelphia, where he enjoyed the reputation of being one of the first operators of his time. Though Dr. Hudson is not credited with any writings relative to the profession, yet he contributed many devices and inventions which make his name dear to those

---

*Dental Advertiser, vol. II., p. 3.
Dental and Oral Science—Dexter, p. 13.

who love the profession.   He died in 1833 at the age of sixty-one.

John Randall* was born in 1773 and graduated at Harvard College in the class of 1802, a class celebrated, for the since displayed, eminent talent of its members.   Randall soon after leaving college began the study of medicine under Dr. John Jefferis of Boston.   After a short period of preceptorship he drifted into the practice of dentistry, and anon established himself in Boston.   The circumstances under which he was lead to turn his attention to dentistry are interesting as showing the character of the practice of that time.   Finding, while at Harvard, that some of his teeth were decaying, he applied to the most prominent dentist of whom he knew and was frankly told by that practitioner that "his business was to put in new teeth" and declined operating, to preserve the natural ones.   This in the light of Randall's general education appeared to be a very limited view of dental science; he at once procured such dental books as he could find and studied them under the impression that disease might and should be remedied without the removal of the diseased member.   His first efforts in dentistry were while at the college, where he not only filled his own teeth but those of such of the students as desired dental attention.   When practicing in Boston

*Dental and Oral Science—Dexter, p. 13.

he was famous as an operator and in extracting teeth was recognized without an equal. The forceps, it is claimed, were used by him long before they were known as an article of merchandise. His success at crowning teeth was very great. He died in 1843 having been an honorable member of the profession and highly regarded by the medical fraternity.

Leonard Koecker * was born in Hanover in 1785 and early in life engaged in merchantile pursuits. At this time he became acquainted with a traveling dentist who took much notice of him, presenting Koecker with a set of instruments. This gift inspired the lad to study dentistry; however, he did not immediately begin the study. In 1807 he came to America, engaged as a commercial agent for an English company, but soon failed, and then turned his attention to dental practice. He located in Philadelphia in 1807 having comparatively little knowledge of the art or science. But his native ability and energy soon overcame any ordinary obstacles. He soon gained both knowledge and practice and in 1822 he stood high in the profession, and did a business of $8,000 per annum, a sum much regarded at that early date. His health, however, compelled a voyage to Europe, and from what we learn, he settled in London, where he was

*Dental and Oral Science—Dexter, p. 15.
Dental Advertiser, vol. II., p. 3.

widely known as an expert dental practitioner. In 1850, then sixty-five years old, he died. Dr. Koecker wrote many valuable articles on dental science, and while in London wrote his famous work on "Dental Surgery." The volume was considered of great merit, and is even to-day quoted by eminent authorities. Some of his methods of operating, especially those on the treatment of the dental pulp, have been almost universally adopted.

Jabez Parkhurst,* born at Newark, N. J., Oct. 4, 1764, after going through a series of occupations, or branches of business, finally settled on dental practice; this at the age of forty-three. He had the advantage of mature judgment, fine address and classical education which secured for him a high literary and social position. After qualifying himself as best he could through the medium of books, he attended lectures on anatomy, physiology and general surgery in his native town. In 1807 he established himself in New York City, where for a period of twenty years he enjoyed the reputation of occupying the highest position in the ranks of the profession, having possibly the largest and most lucrative practice in that part of the Union.

Such in brief is the sketch of the pioneers of the profession in this country. Our study of these characters demonstrates that the art and science

*Dental Advertiser, vol. II., p. 3.

was nurtured by noble students; that all through the primitive years of dentistry there have been men of education and mechanical ingenuity wielding a controlling power, and who had a just appreciation of the needs of dental practitioners. Many more gentlemen of later date and not less strong claims for a biographical notice might be here included; but will postpone speaking of them and let their acts tell for them as embodied in the pages to come, rather than mention them specially. In thus deciding no negligence of their claims or depreciation of their worth and ability is intended, and trust that fair consideration may be shown them.

The following written by Dr. Welch, the able editor of the "Items of Interest," speaks appreciative words of the fathers of our profession:*

"Let us not ignore the fathers of our profession. They worked under great disadvantages, yet many became shining lights. When we were groping our way in darkness they lent us their light, and showed us the way to success. We rejoiced in their light until we too became illuminated and went forth to attract and conquer. Others are silently passing over the hill toward the setting sun. As we mark their stately steps, the glow of their evening sky brightens our faces and gives us inspiration. We walk in the highway they made

*Items of Interest, vol. XIV., p. 255.

OF DENTAL PROSTHESIS. 187

for us; let us gratefully acknowledge their services and still further improve and beautify the way, and leave it to our successors as a kingly-way, dignified and honorable.

"As one and another of our bright lights passes away, we are too apt to believe there are fewer left by their departure. There is a bright side to this picture. Others are coming with torches brightly burning, brighter than those of their predecessors. They are moving up and making the way beautiful. Give them room; give them cheer; while we very properly sing requiems to our honored dead, let us not refuse merited encomiums to the noble living. We have still leaders who are giants. They are the world's grand men. When we lay them, one by one, into the cold grave, we will acknowledge their merits; let us acknowledge them now, while they live and can hear it, bow to their wisdom, learn of their skill, sit at their feet as humble disciples, and let them see we appreciate them.

"Our best teachers would be better if we gave them greater evidence of our appreciation. Even when such are specially invited to our gatherings, they are used more as gold-headed canes to adorn the meetings than as illuminators; and instead of generously recompensing them, the officers think they are doing wonders if they pay their bare expenses."

From the time of Dr. Greenwood's successful appliance for Gen. George Washington until 1820, successors from Europe and adoption of the profession by native born Americans greatly increased the number of dentists. The darkness which shrouded scientific dentistry in the seventeenth century was now dispelled, and in the beginning of the nineteenth century, with the development of a new political life in the American empire, saw dentistry given a professional and social standing worthy of all its importance in alleviating the woes of the human family.

A curious specimen of medieval American prosthetic art is in the custody of Dr. Brophy. The piece consists of ivory, carved to fit the upper and lower jaws. The block was carved so as to fit the alveolar-ridge or process, and on each side a bicuspid and two molars were carved in the same block. The anterior teeth consist of human teeth fastened with gold to the ivory. The carving was skillfully done, the sulci and cusps of the molars were artificially reproduced. The ivory, however, did not resist the actions of the fluids of the mouth, and thus the ivory was attacked by caries similar to the effect of that disease on the natural teeth. Another unique specimen of early American dental skill is in the possession of the H. D. Justi Dental Manufacturing Company, and they have kindly allowed the cases to my table that I might describe the crude workmanship.

The first case, an upper partial gold plate supplying all the teeth except the two twelve-year molars, is a grand success and would do credit even to to-day. The gold plate snugly fitted the alveolar process and slightly covered the palatal portion of the mouth. The teeth which were carved from the tusks of the walrus were ingeniously shaped, and the gold pins which penetrated the teeth from cutting edge through the main shaft and body, penetrated the gold plate and were soldered to the latter on the palatal surface. The spacing between the teeth so closely approximates nature that the wearer of this denture could use the tooth-pick to good advantage. The plate, instead of clumsily enveloping the two natural molars, has skillfully fitted gold clasps which materially aided in keeping the denture properly positioned.

The second case, a partial upper and lower connected by means of spiral-springs, is an intricate arrangement but demonstrates to a dot that the primitive dental practitioners of America were of an ingenious order. The upper partial denture supplied the upper labial teeth, while the two twelve-year natural molars were still in position. The artificial substitutes were of two kinds, the anterior six teeth were carved from bone and attached with gold, similar to modern methods; the bicuspids and molars, instead of being carved individually, were executed in one solid block, with cross groves on the cutting surface to represent the sulci of the natural teeth and the entire block being rivited to the gold plate which simply followed the alveolar ridge. Around the natural molars were clasped two gold-bands. The lower denture was attached to the upper one with gold spiral-springs. This denture, like the upper, was partial, and intended to supply substitutes for the central incisors, one bicuspid on the right and two molars on the left side of the jaw. The two central substitutes were human teeth imbedded in sockets of-gold which received the respective roots of the two substitutes. From the centrals there was a continued bar of platinum, ending in clasps to surround the lateral and bicuspid natural teeth. The artificial molars and bicuspids were made of one solid block of bone and lines of demarcation, representing the divisional spaces of the teeth. The natural molar on the right side was encircled by a platinum clasp.

No matter how unique or ungraceful their first attempts after their ideal; rising from step to step with progressive thought, ever keeping in view that philosophical principle that he is the greatest philanthropist who helps the greatest number, being followed from time to time and succeeded by others, and from a congress of thought each new idea was alike heralded to all, keeping no secrets but ever extending the olive-branch of fraternal peace and kindly greeting.   The public becoming aware, by benefits received of its indispensibility, gave to them encouragement and increased patronage, and the practicability of a dental college was discussed.

Enough at least has been given which conclusively proves that dental science, though, perhaps, rude at first, is not of very recent origin.

No matter by whom, when, or by what means it first became a thought and opened up its petals to receive the warmth and genial sunshine of a credulous people, it stands to-day one of the first and noblest of the sciences, extending and taking in its embrace almost every part of every continent and inhabited islands of the seas, and with it all pertaining to its first grand idea—that of doing good to humanity.   But not until the eighteenth century did dentistry become the subject of much critical inquiry and thorough investigation.   Men of education and talent devoted themselves to it exclusively, and from that period it has progressed rapidly in importance; and within this brief period its progress has been so rapid that to understand and expound the whole extent of the subject is already far beyond the possible.

To the American dentists is due the glory of establishing nearly all the strides of dental advancement and placing the profession in a commanding position. The mechanical devices, and various labor-saving appliances and materials contributed by them to the advancement of dentistry, are numerous and important, the mere specification of which would consume hours of time.

Perhaps the most important American donation to dental art is the production of porcelain teeth. This industry was an outgrowth of Dental Prosthesis. Though of French origin, their perfection is due entirely to the untiring efforts of the American manufacturers.

To Drs. A. A. Plantou and C. W. Peale * of Philadelphia must be awarded the credit of manufacturing, in 1820, the first porcelain teeth in the country; but S. W. Stockton of Philadelphia and James Alcock of New York in 1850 began their production upon a more extended scale, for the purpose of supplying the profession at large, and thus initiated an industry which has gained remarkable proportions. The present degree of perfection in moulding and enameling the teeth was not attained until some years later, nor was the color so life-like, or the shades so varied. For many years after the introduction of porcelain

---

*He was the son of the famous Rambrandt Peale, and C. W. Peale is to be remembered, aside from his attainments in Dental Prosthesis, for his grand display of genius in the painting of an authoritative portrait of Gen. George Washington. C. W. Peale was a universal genius. He was successively a saddler, silversmith, watchmaker, dentist and portrait-painter, in which latter vocation he acquired his most enduring reputation.

teeth the best artists were unable to make them
sufficiently perfect in form and in color to give
good appearance in the mouth. The porcelain
teeth were indestructible by the oral secretions,
while the then popular carved blocks of ivory
decayed, and became offensive and were eventually
injurious. This latter difficulty with the ivory
teeth rendered the porcelain teeth very desirable,
and in consequence improvement in their manufacture continued. Americans have been very
successful in this form of advance, and mainly
instrumental in bringing the manufacture to its
present high state of perfection. So far, indeed,
has this art of tooth production from felspar, silex
and kaolin been developed that any dental form
and coloring desirable can be rendered so as to
deceive any but the trained eye of an expert.
These improvements in the fabrication of porcelain
teeth which have so admirably displayed the possibilities of the manufacturers in the transparency of
the tooth, the granular appearance and flesh-
like tint of the gums, and the unlimited shades
were due to the persistency of Dr. Elias Wildman
of Philadelphia, who began his numerous experiments in 1837.

In 1844 Samuel S. White, a nephew of Samuel W. Stockton, began the production of these
teeth in Philadelphia, and this was the initiatory step in an enterprise which has since

grown to be the largest of its kind in the world.

Numerous improvements are accredited to Mr. S. S. White, but not all the comment does he reap, since shortly after his attempts and experiments H. D. Justi of Philadelphia, believing that he, too, saw an opportunity for improving the aesthetic effects of porcelain teeth laboriously toiled to reach the goal of perfection in the yet crude art.

Dentistry seemed destined to rise as a learned science and profession; and able artists in all parts of our country lent a helping hand to establish the truths of dental science; and among the many depots, shops and factories which have aided in this good work we find those of Johnson, Lund & Co., Gideon Sibley, William M. Speakman, The Harvard Dental Company; more recently the Wilmington Dental Manufacturing Company, and various others have loomed above the horizon, and these, too, are companies whose artists are highly skilled in ceramic work.

The various dental companies are now producing all the various appliances, instruments and materials; it is estimated that not less than 100,000 teeth a month, or nearly 12,000,000 per annum are manufactured in America alone.

All this present perfection and completion has not been brought to us in a single birth, but, on the contrary, required time, patience, talent and expense.

About the same time that the several dental companies were devoting their attention to the newly discovered art, the individual dentists, too, were busy calculating on the self-same subject. The result was that early in 1850 Dr. John Allen, a distinguished dental practitioner, devised a method embracing original and important modifications in the aesthetic shapes, colors and arrangement of the dental substitutes. The exactness with which Dr. Allen represented the natural gum-tissue gave to his process and discovery the name of Continuous Gum work. The intimate but later identification of Drs. Hunter and Haskell with continuous gum work have rendered their names familiar, as being skilled and devoted to this specialty of Dental Prosthesis; and their respective contributions to the development of a perfect process in the departure has done much towards establishing for the dental profession a lasting glory.

The announcement in 1851 of Nelson Goodyear's invention for making the hard-rubber compound, substantially termed "Vulcanite," turned the attention of those interested in the manufacture of various small articles for use and ornament to the adoption of this material which was announced as a substitute for horn, bone and ivory, susceptible of being colored and possessing the plasticity of gutta-percha, while it was exempt from the actions of heat, cold and acids.

In 1855 the first patent was obtained for making a dental plate in hard rubber. The introduction of vulcanite into the profession materially injured the general tone of dexterity and science among the dentists; the ease with which vulcanite is worked invited many into the profession who were utterly unfit and incapable of scientifically replacing the lost organs. Prior to the Goodyear invention the dentist was obliged to be more than a mere mechanic, since the various metals were then used as bases for dentures.

Celluloid, like vulcanized rubber, a cheap base for artificial dentures, was first introduced in 1869, and during the existence of the "rubber patents" was much used by those who objected to becoming licensees of the Goodyear Rubber Company. The advantages claimed for the celluloid were its unlimited artistic possibilities, resemblance in color to the natural tissues, readily tolerated by mucous membrane, elasticity under strain, adaptability for partial or complete dentures and the readiness with which it could be applied to the correction or concealment of all oral deformities.

But celluloid like, we hope, rubber, too, "has seen its days of triumph," and every dentist who has the welfare of his profession at heart is pronounced in his desire to see all cheap and injurious plastic base-dentures shelved, thus to make room for the furtherance of our acquaintance with metals, minerals and numerous other materials for dental plates.

An incidental but most important advantage to dentistry, accompanying the revival of gold-crown and bridge-work, is the requirement of increased manipulative and artistic skill on the part of the operator. The character of a large proportion of the vulcanite work of the last few years has a sulphurous oder about it suggestive of the adage, "facilis est descensus Averni." To such a facile method of constructing artificial dentures is due the advent of a class of dentists who have "picked up the business" in a few months of untutored experiment. To clumsy, disfiguring dentures, so bulky as to impair speech, and so incompletely finished as to occasion sore mouths, have been chargeable in great degree the discomforts and diseases attributed to the cheaper base-dentures.

The increasing demand for dentures on gold bases must of necessity raise the standard of qualification for dental practice. For the promotion of this most desirable end the dental society clinics, increasing as they are, alike in frequency and in interest, have become potent factors.

If you possess the requisite skill, and will do yourself the justice to use it, you can make even a rubber plate that will not disgrace you, and get paid for it, too. While visiting different offices I have often been amused at the assumed airs: ''Oh, I never dirty my hands with that class of work, I leave that to the cheap Johns, or, as they say down South, 'I have a nigger to do my plate work.'" Occasionally you find some old in the prefession who delegate the "mechanical to the shops." Such almost invariably demonstrates the fact that they are not capable of constructing a perfect artificial case. The renowned Josephs of San Francisco said: "After twenty years' experience, the first thing I would impress upon a tyro is that fiddle-making is a trade but violin-making is an art." Tooth-filling as well as plate-making is too often a mere trade, but properly restoring loss with artificial teeth, with all that is implied in the operation, is the acme of dental art if there is any art in dentistry.

To-day the vast variety in shape, size, color, etc., of the porcelain teeth, gives opportunity for the selection of forms suitable to nearly every case which presents itself to the general practitioners. The assortment must of necessity be very large and varied to meet the wants of the prosthetic dentist.

Porcelain is a material in which the beauty of the result well repays the highest exercise of art. It has been for centuries the favorite material for expressing the poetry of form. To the famous Etrurian vases of antiquity, the exquisite gems of the

majolica of the sixteenth century may be named
in proof of the fitness of pocelain to embody the
conception of genius.

Dental porcelain is worthy of such associations;
not only like these does it delight the eye and give
evidence of high aesthetic cultivation, but it adds
to beauty, the charm of usefulness.

It is customary to attribute the rapid growth of
dental art, since 1840, to dental associations, colleges, journals and its didactic literature—and
worth much truth. But to porcelain it owes its
very existence as an aesthetic art, and the largest
extent and utility as a prosthetic science. It was
altogether impossible for perishable human teeth,
or their wretched imitations in ivory, to offer such
tempting fac-similes of nature as we meet in proce-
lain productions.

The dental depots not only rendered service by
superior excellence of surgical instruments, and
prosthetic appliances and materials, but they directly benefited the science and art of dentistry by
releasing the practitioners from the manufacturing
toil, and gave them time for the acquirement of increased knowledge and skill in the various departments of dental science.

Few people can comprehend, and fewer still
thoroughly appreciate, the many noble favors the
dental profession deals out to suffering humanity. Among the myriads of grand achievements

wrought, very few excel the prosthetic accomplishments in the cure and relief of deformed palatine
organs, or what is known to surgeons by the name
of cleft-palate.

This is one of the most distressing deformities
to which the human frame is liable. The unfortunate sufferer is compelled, in a great measure, to
be an alien among his fellow creatures; an object
of compassion to the considerate, he is often made
painfully conscious of his deformity by heartless
companions. And were he gifted with eloquence
of Demosthenes or Webster he could make little
more use of his endowments than a mute. Fortunately this painful defect, which may be either
accidental or congenial, is no longer reckoned one
of the incurable, since Dental Prosthesis has risen
to the present pinacle of perfection. The same is
true with reference to irregular teeth and malfor-
mations of the jaw in general.

Probably no feature in the annals of dentistry
is worthy of so much comment and merits such
·deep consideration as the subject of crown and
bridge-work. For the past few years it has been
a subject of overwhelming magnitude, and to-day
is the tidal wave in Dental Prosthesis, stirring
to eloquence its admirers and agitating the dental
fraternity in an effort to establish its worth or use-
lessness.

In searching for the history of crown-work we

are utterly dumfounded, and upon becoming cognizable with its antiquity we are almost prompted to chronicle it as a lost art recklessly omitted in the list of ancient advertisements. The operation of pivoting teeth, synonomous to our modern crowns, is one of the primitive methods of replacing lost natural ones, and it is indeed debatable whether the plate preceded the pivot or vice versa.

Historical theses unfortunately left but few "foot-prints on the sands of time," and in consequence we are much at loss as to the data of the subject. The artificial replacement of the loss of a portion of the teeth by crowns was first written about Robert Woofendale in 1783, when he wrote regarding the act of joining artificial crowns to the roots of natural teeth; he tells that where the enamel had suffered severe destruction it was advisable to cut off the injured remaining portion of same and replace by means of an artificial one, and he dwelt elaborately on the wonderful stability of such a piece of work. It is extremely difficult, indeed, almost impossible to arrive at anything like certainty in determining priority of invention or introduction of any improvements earlier than 1839, at which date recorded history of dentistry was inaugurated by a dental journal.

Statements are made pro and con relative to crown-work and its early popularity. Scrap den-

tal literature bears witness that in 1804 current
newspapers contained lengthy articles on the "art
of restoring roots of teeth by means of wire-pivots,
wooden-pivots, cotton-wrapped-pivots and screw-
pivots."

I deem it unnecessary to enter into a detailed
description of the numerous new and improved
modern crowns, and it would be an extensive work
that would treat of each and every method of construction.   Suffice it to say that Drs. Lawrence,
Foster, Bean, Richardson, Buttner, Thomas, Leech,
Webb, Williams, Hay, Boice, Weston, Carman,
Hunter, Bonwill, Davis, How, Logan, Richmond,
Land and Evans have contributed a commendable
amount towards bringing the artificial crown within
the rays of perfection.

True, we have seen marked advancements in
the construction and appliance of the various
crown substitutes, and bow in all homage to pay
tribute to those great minds who conceived the
grand and satisfactory results.   Shall we honor
most the practitioners who first wrought the happy
means or who later perfected it, is one of those
questions which shall be interesting grounds for
debate.

As to the history of bridge-work we are about
as knowing as can be expected on account of existing circumstances; in brief, the developments of
this departure are not unlike those of crown-work,

and the same catalogue of dentists' names adorn
its tabular records, with the exception that the
names of Drs. Bing, Case, Brown, Starr, Ludwig
and Melotte are directly connected with the numerous steps of improvement.

On the system of bridge-work Dr. Bennett
makes these few and worthy remarks:*

"Until within a few years dentistry consisted
chiefly of fillings and plates, neither of which have
always met the requirements in every case. During the last decade there has been rapidly developed what may be properly called a third branch
of dentistry, closely related to the preceding,
which has been practiced with varying degrees of
success and failure, and has been advocated and
opposed with almost all degrees of knowledge and
prejudice. This new system, which is, strictly
speaking, an old system revived and greatly improved, is called crown and bridge-work. Though
resembling dental prosthesis more than operative,
this mode of restoring lost organs partakes of the
nature of both, and begins where filling ends.
Filling by restoring lost tissue attempts to save the
crown, while crowning may be said incidentally at
least to save the root. If it can be shown that
this new system of prosthesis has a scientific basis
and conforms to correct mechanical principles,
time and practice can be depended on to devolop

---

*Dental Cosmos, vol. XIX., June.

skill in the art of construction. Of the few things we know with certainty about dentistry, one of the first is that any mechanism that ignores the facts of science, or does not comply with conditions and meet requirements, must inevitably fail.

"Though I think that this new system of substitution has been well named, yet the term bridge-work is associated in many minds with all that is discreditable to the dentist and disgusting to the patient. While not strictly correct in all cases, this new method might not inaptly be termed 'continuous crown-work'—a name that perhaps would weaken some of the opposition and prejudice that bridge-work has encountered."

DENTAL EDUCATION.*

It was not until 1839 that any movement in the way of organization was made on the part of American dentists to elevate their profession to a strictly educational basis. In view of uniting the widely separated members of the profession, a medium was established in the form of the "American Journal and Library of Dental Science." The journal was published in Baltimore and ably edited

---

*The author would recommend for the reader's consideration the following valuable papers on "dental education":

Trans. of Am. Dent. Associ., 1891-1892, Sec. II., by L. Ottofy, D. D. S.
Dental Review, vol. IV., p. 664, by C. N. Pierce, D. D. S.
Ibid, p. 672, by W. H. Atkinson, M. D., D. D. S.
Ibid, p. 677, by A. H, Thompson, D. D. S.
Ibid, p. 692, by R. Ottolengui, D. D. S.
Ibid, vol. V., p. 144, by C. N. Johnson, D. D. S.
Dental Cosmos, vol. XXXIV., p. 92, by C. E. Bentley, D. D. S.

by Chapin A. Harris and Eleazer Parmly. In connection with this movement it was the ambition of
Dr. Harris to organize a dental school as an ad-
junct to the medical department of the University
of Maryland. The practice of dentistry at this
time, however, being with few exceptions at a very
low ebb, the faculty of the university rejected the
proposition of Dr. Harris, they giving as an excuse,
"that the subject of dentistry was of little consequence, and thus justified their unfavorable action."
The rejection seemed to give Dr. Harris new
energy and stimulated in him a new desire, and as
a result the Baltimore College of Dental Surgery
was established.

The origination of institutions for the teaching
of dentistry was due to the persistent and determined efforts of a few men of markedly liberal views
of the profession. The great necessity for some
radical change in the method of imparting dental
instruction was sufficiently evident to any observ-
ing and impartial mind long before 1840. The
practice of taking private students was then everywhere in vogue, and little could be said in its disfavor so long as the studentship was properly conducted and the teacher imparted to his disciple
that which he himself knew on the subject; and
danger in this direction could not be apprehended
when certain men of established reputation were
the teachers. But such, unfortunately, was not

always the case. The eminent dentists of that period charged extremely high prices for studentship. Dr. Parmly offered to "receive students and to render them fit for practice for one thousand dollars."* And other dentists of reputation asked similar fees. Such high charges were of themselves sufficient to debar the great majority from seeking dental education, and consequently were either forced to enter the profession without knowledge of the art or science, or pay in advance a fee which in those days meant a fortune. This state of affairs was naturally viewed with regret by such as were above professional jealousies and secrecy in methods which then almost universally prevailed. Liberal dentists strove to induce a change in this condition of things. The quickest and most surely effectual methods to produce the desired result appeared in establishing a dental college.

The labors of that noble man, Chapin A. Harris, whose name should be all the more honored for having devoted his time and his talents less to the accumulation of wealth than to the development of dentistry into a scientific profession, together with his associates, lifted the practice of dentistry from that obscure empiricism which had largely characterized it into the clear light of scientific inquiry and demonstration, by laying that foundation in

the Baltimore College of Dental Surgery, upon
which our system of dental education has been
built.

So long as the various processes of dental
surgery and laboratory continued to be held as
carefully guarded secrets which might be filched
from one by his neighbor, engendering a narrow-
minded jealousy of every means and appliance of
which one had not the monopoly, so long there
could be neither professional character nor standard; and to the disciple of such a school it was
enough to be as his master. But the leaven of the
old order of things began to work. The aspiration for better instruction, necessary to the attainment of a professional standing, began to imbue
the minds of younger practitioners, encouraged by
the success of their seniors and incited by the
desire of popular approval.

The motives of Dr. Harris and his associates
in this undertaking are set forth in Dr. Harris'
introductory lecture, delivered on the occasion of
the opening of the college, in the year 1840. Of
the state of dentistry at this time Dr. Harris says:

"No credential or evidence of competency having been looked for or required, the profession has
become crowded with individuals ignorant alike of
its theory and practice; and hence its character
for respectability and usefulness has suffered in
public estimation, and a reproach has been

brought upon it which it would not otherwise have deserved.

"The community is at present unable to discriminate between the well-educated and skillful practitioner and the merest pretender—and until it shall be able to do this more or less of the reproach that is brought upon the pursuit by the latter will be visited upon the former.

"Accessible as has been the calling of the dentist to all that were disposed to engage in it, and that, too, without regard to qualification, it has been resorted to by the ignorant and illiterate, and, I am sorry to say, in too many instances, by unprincipled individuals, until it now numbers in the United States about twelve hundred, and of which I think it may be safely asserted that not more than one-sixth possess any just claims to a correct or thorough knowledge of the pursuit. That, under such circumstances, it should occupy a place in the world's estimation inferior to that to which it would otherwise be entitled, is not a subject of wonder.

"Dentistry" he characterizes as "a most useful and valuable department of medicine;" as a "branch of surgery.

"Of the qualifications necessary to be possessed by a dental practitioner, and the time required for their acquisition, few seem to be aware. On this subject an erroneous opinion

seems pretty generally to prevail. A little mechanical tact or dexterity is thought by some to
be all that is requisite to a practitioner of dental
surgery, and that this could be obtained in, at
most, a few weeks. The prevalence of this belief
has given countenance to the assumption of the
profession by individuals totally disqualified to
take upon themselves the exercise of its compli-
cated and difficult duties. But it is to be hoped
that the day is not remote when it will be required
of those to whom this department of surgery shall
be intrusted to be educated men, and well instructed in its theoretical and practical principles.
Elevate the standard of the qualifications of the
dental surgeon to a level with those of a medical
practitioner and the results of his practice will be
always beneficial, which at present are frequently
the reverse. Require of the practitioner of dental
surgery to be educated in the collateral sciences
of anatomy and physiology, surgery, pathology
and therapeutics, and the sphere of his usefulness
and his respectability will be increased."

With these views the faculty entered upon
their work, with earnest promises of faithfulness
to the important trust they had assumed. "Aware,"
said Professor Harris, "of the responsibility
that rests upon them, the faculty will spare no
efforts to make it creditable to the State that
created it and beneficial to the public. Conscious

that its claims to respectability and usefulness will depend upon the manner in which they shall discharge their duties, it will .be their constant endeavor to impart not only correct, but thorough, theoretical and practical information; persuaded that without this it is impossible for any to practice the art with credit to themselves, or for the benefit of their employers, they are resolved to admit none to the honors of the institution except such as possess it.   In short, they are determined that no reproach shall rest upon them for fixing a standard of qualification that shall not at once be respectable, and entitle those coming up to it to the confidence of an enlightened community."

Dr. Harris' proposition was a bold one, requiring a degree of courage not possessed by the average man.   It was an untried field; the prospects for emoluments for the pioneers were not inviting, and failure would engulf all concerned in irremediable ruin.

In the first scholastic class there were two graduates, Robert Arthur and R. Covington Mackall. Thus in due time the college labeled with dignity and honor a small class as "Doctors of Dental Surgery."

Thus a system of education was initiated which immediately placed the practitioners of dentistry upon an equal footing with other liberal professions.   All hail the banner of the old Baltimore

College of Dental Surgery! the progenitor of much good and the Alma Mater of Alma Maters, claiming among her collegiate alumni your adopted mother.

Hence, with this college as the nucleus, many prototypes have since been generated, and the good effect these various institutions have on the public none can more sincerely testify than suffering humanity. The profession as well as its many faithful representatives have steadily and surely risen, never again to fall.

As a matter of history we append the names of all doctors who have filled the chair of Dean, Professor of Operative Dentistry, and Professor of Dental Prosthesis. In the Baltimore College of Dental Surgery:* Deans—Chapin A. Harris, M. D., 1840-1841; Thomas E. Bond, M. D., 1841-1842; Washington R. Handy, M. D., 1842-1853; Philip H. Austen, M. D., 1853-1865; F. J. S. Gorgas, M. D., D. D. S., 1865-1882; Richard B. Winder, M. D., D. D. S., 1882 ——. Professors of Operative Dentistry—Chapin A. Harris, M. D., 1840-1846; Amos Westcoat, M. D., D. D. S., 1846-1849; C. O. Cone, M. D., 1849-1852; Alfred A. Bandy, M. D., 1852-1856; Edward Maynard, M. D., 1856-1860; F. J. S. Gorgas, M. D., D. D. S., 1860-1879; Richard B. Winder, M. D., D. D. S., 1879-1884; John C. Coyle, D. D. S., 1884-1888;

*Informed by Dr. R. B. Winder.

Richard B. Winder, M. D., D. D. S., 1888 ———.
Professors of Dental Prosthesis—Chapin A. Harris,
M. D., 1840-1846; Amos Westcoat, M. D.,
D. D. S., 1846-1849; C. O. Cone, M. D., 1849-
1852; Philip H. Austen, M. D., 1852-1873; James
B. Hodgkin, D. D. S., 1873-1889; William B.
Finney, D. D. S., 1889 ———.

The Ohio Dental College* of Cincinnati was
chartered in 1845 and began its first session in
November of the same year.   Deans—Jesse W.
Cook, M. D., D. D. S.; J. Taft, D. D. S.; H. A.
Smith, D. D. S.   Professors of Operative Dentistry
—James Taylor, D. D. S.; J. Taft, D. D. S.; H.
A. Smith, D. D. S.   Professors of Dental Pros-
thesis—James Taylor, D. D. S.; W. H. Hunter,
D. D. S.; John Allen, D. D. S.; H. R. Smith, D.
D. S.; J. R. Clayton, D. D. S.; Frank Bell, D. D.
S.; W. Van Antwerp, D. D. S.; Grant Molly-
neaux, D. D. S.

The Philadelphia College of Dental Surgery†
was the first college in Philadelphia and was chartered in 1850.   The school had four sessions and
in 1856 ceased to exist.   Dean—Elisha Townsend,
M. D., D. D. S.   Professor of Operative Dentistry
—Elisha Townsend, M. D., D. D. S.   Professor
of Dental Prosthesis—T. L. Buckingham, M. D.

The New York College of Dental Surgery‡

---

*Informed by Dr. H. A. Smith.
†Dental and Oral Science, Dexter, p. 184
‡Ibid, p. 183.

opened in Syracuse, N. Y., and chartered by the State in 1852.* Dean—A. Westcoat, M. D. Professor of Operative Dentistry—W. Dalrymple, D. D. S. Professor of Dental Prosthesis—W. Dalrymple, D. D. S.

The Pennsylvania College of Dental Surgery† of Philadelphia was chartered in 1856. It was the result of the downfall of the old Philadelphia College of Dental Surgery. Deans—Elisha Townsend, M. D., D. D. S.; Thomas L. Buchingham, M. D.; William Calvert, D. D. S.; C. N. Pierce, D. D. S. (eight years); Elias Wildman, M. D., D. D. S.; Charles Essig, M. D., D. D. S.; C. N. Pierce, D. D. S. (sixteen years). Professors of Operative Dentistry—Elisha Townsend, M. D., D. D. S.; J. H. McQuillen, M. D., D. D. S.; C. N. Pierce, D. D. S.; James Truman, D. D. S.; E. T. Darby, M. D., D. D. S.; C. N. Pierce, D. D. S. Professors of Dental Prosthesis—T. L. Buckingham, M. D.; William Calvert, D. D. S.; Elias Wildman, M. D., D. D. S.; Charles Essig, M. D., D. D. S.; Wilbur F. Litch, M. D., D. D. S.

The Philadelphia Dental College‡ was chartered in 1863 and is located in Philadelphia. Deans—J. H. McQuillen, M. D., D. D. S., 1863-1879; D. D. Smith, M. D., D. D. S., 1879-1880; J. E. Garret-

---

*Became defunct in 1855.
†Informed by Dr. C. N. Pierce.
‡Informed by Dr. T. C. Stellwagen.

son, M. D., D. D. S., 1880 ——. Professors of Operative Dentistry—C. A. Kingsbury, D. D. S., 1863-1865; George W. Ellis, D. D. S., 1865-1866; Thomas C. Stellwagen, M. D., D. D. S., 1869-1879; D. D. Smith, D. D. S., 1879-1881; S. H. Guilford, D. D. S., Ph. D., 1881 ——. Professors of Dental Prosthesis—Thomas Wardle, D. D. S., 1863-1867; D. D. Smith, D. D. S., 1867-1881; S. H. Guilford, D. D. S., Ph. D., 1881 ——.

The New York College of Dentistry* was chartered in 1865 and began its first session in New York City in November, 1866. Deans—Norman W. Kingsley, 1866-1869; Frank Abbott, M. D., 1869 ——. Professors of Operative Dentistry—W. H. Dwinelle, M. D., D. D. S., 1866-1867; Edward Dunning, 1867-1868; Frank Abbott, M. D., 1868 ——. Professors of Dental Prosthesis—Norman W. Kingsley, 1866-1869; C. A. Woodward, 1869-1873; C. A. Marvin, D. D. S., 1873-1877; J. Bond Littig, D. D. S., 1879 ——.

The Missouri Dental College† was chartered in 1866 and is located in St. Louis. Deans—Homer Judd, M. D., 1866-1874; C. W. Rivers, D. D. S., 1874-1875; W. H. Eames, D. D. S., 1875-1878; H. H. Mudd, M. D., 1878 ——. Professors of Operative Dentistry—H. E. Peebles, D. D. S., 1866-1869; H. S. Chase, M. D., D. D. S., 1869-

1874; C. W. Rivers, D. D. S., 1874-1875; I.
Forbes, D. D. S., 1875-1877; Homer Judd, M. D.,
1877-1878; J. Ward Hall, D. D. S., 1878-1879;
A. H. Fuller, M. D., D. D. S., 1879 ——. Professors of Dental Prosthesis—W. H. Eames, D. D.
S., 1866-1875; A. H. Fuller, M. D., D. D. S.,
1875-1876; M. A. Bartleson, D. D. S., 1876-1877;
H. H. Keith, D. D. S., 1877-1880; W. N. Mor-
rison, D. D. S., 1880-1884; J. G. Harper, D. D. S.,
1884 ——.

The New Orleans Dental College* was chartered
in 1861 and ceased to exist in 1878. Deans—John
S. Clark, D. D. S.; Andrew F. McLain, M. D.,
D. D. S. Professors of Operative Dentistry—A.
F. McLain, M. D., D. D. S.; W. F. Chandler, D.
D. S. Professors of Dental Prosthesis—G. J.
Fredrichs, D. D. S.; Charles E. Kells, D. D. S.

The Dental School of Harvard University† was
chartered in 1867 and the school is located in
Boston. Deans—Nathan C. Keep, M. D.; Thomas
B. Hitchcock, M. D.; Thomas H. Chandler, A. M.,
D. M. D. Professors of Operative Dentistry—
George T. Moffatt, M. D.; L. D. Shepard, D. D.
S.; Thomas Fillebrown, M. D., D. M. D. Profes-
sors of Dental Prosthesis—Nathan C. Keep, M. D.;
Thomas H. Chandler, A. M., D. M. D.

The Boston Dental College‡ was chartered in

*Became defunct in 1877.
†Informed by Dr. T. H. Chandler.
‡Informed by Dr. J. A. Follett.

1868 and is located in Boston. Dean—J. A. Follett, M. D. Professors of Operative Dentistry—
I. J. Wetherbee, D. D. S.; William Baker, D. D.
S.; J. H. Daly, D. D. S. Professors of Dental
Prosthesis—S. J. McDougall, M. D.; H. F. Bishop,
D. D. S.; N. N. Noyes, D. D. S.; A. Lawrence,
M. D., D. D. S.; J. K. Knight, D. D. S.

The Maryland Dental College* was founded in
1873 and is located in Baltimore. Deans—R. B.
Winder, M. D., D. D. S.; F. J. S. Gorgas, M. D.,
D. D. S. Professors of Operative Dentistry—E.
P. Keech, D. D. S.; H. G. Ulrich, D. D. S.; F.
J. S. Gorgas, M. D., D. D. S. Professor of Dental
Prosthesis, F. J. S. Gorgas, M. D., D. D. S.

The Dental College,† University of Michigan,
was instituted in 1875, and established in Ann
Arbor. Dean—J. Taft, D. D. S. Professors of
Operative Dentistry—J. Taft, D. D. S., and J. A.
Watling, D. D. S. Professors of Dental Prosthesis—J. A. Watling, D. D. S.; N. S. Hoff, D. D. S.

The Indiana Dental College‡ was chartered in
1878 and is located in Indianapolis. Secretaries—
J. E. Cravens, 1879-1889; J. N. Hurty, M. D.,
Ph. D., 1889-1892; J. E. Cravens, 1892 ——.
Professors of Operative Dentistry—J. E. Cravens,
1879-1889; J. R. Clayton, D. D. S., 1889-1891;
J. E. Cravens, 1891 ——. Professors of Dental

*Dental and Oral Science—Dexter, p. 194.
†Informed by Dr. J. Taft.
‡Informed by Dr. J. E. Cravens.

Prosthesis—Joseph Richardson, M. D., D. D. S., 1879-1882; T. S. Hacker, D. D. S., 1882-1892; E. E. Reese, D. D. S., 1892-1893; R. T. Oliver, D. D. S., 1893 ——.

Dental Department of the Pennsylvania University* was chartered in 1878 and is located in Philadelphia. Deans—Charles J. Essig, M. D., D. D. S., 1878-1882; James Truman, 1882 ——. Professors of Operative Dentistry—Edwin J. Darby, M. D., D. D. S., 1878 ——. Professor of Dental Prosthesis—Charles J. Essig, M. D., D. D. S., 1878 ——.

The Dental Department of Vanderbilt University† was organized in 1879 and is located at Nashville, Tenn. Dean—W. H. Morgan, M. D., D. D. S.; Professors of Operative Dentistry—J. C. Ross, D. D. S.; H. W. Morgan, M. D., D. D. S. Professor of Dental Prosthesis—R. R. Freeman, M. D., D. D. S.

The Chicago College of Dental Surgery‡ was chartered in 1882, and was the first dental college established in Chicago. Dean—Truman W. Brophy, 1882 ——. Professors of Operative Dentistry—G. H. Cushing, D. D. S., 1882-1889; Edmund Noyes, D. D. S., 1889-1890; C. N. Johnson, L.D. S., D. D. S., 1890 ——. Professors of

*Informed by Dr. James Truman.
†Informed by Dr. W. H. Morgan.
‡Informed by Dr. Truman W. Brophy.
Dental Department of Lake Forest University.

Dental Prosthesis—L. P. Haskell, 1882-1886; W. B. Ames, D. D. S., 1886-1890; C. S. Case, M. D., D. D. S., 1890 ——.

The Dental Department of Iowa University was chartered in 1882 and is located in Iowa. Deans—L. C. Ingersoll, A. M., D. D. S.; A. O. Hunt, D. D. S. Professor of Operative Dentistry —W. O. Kulp, D. D. S. Professor of Dental Prosthesis—A. O. Hunt, D. D. S.

The Dental Department of Maryland University* was founded in 1882 and is located in Baltimore. Dean—F. J. S. Gorgas, A. M., M. D., D. D. S. Professor of Operative Dentistry—F. S. A. Gorgas, A. M., M. D., D. D. S. Professor of Dental Prosthesis—James H. Harris, M. D., D. D. S.

The Northwestern College of Dental Surgery† was chartered in 1885 and is located in Chicago. Deans—I. Clendenen, D. D. S., 1885-1886; J. F. Austin, D. D. S., 1886-1887; Joseph Haven, M. D., 1887-1889; B. Grant Jefferies, M. D., Ph. G., 1889-1890; R. W. Clarkson, A. M., D. D. S., 1890-1891; B. Grant Jefferies, Ph. G., M. D., C. M., 1891 ——. Professors of Operative Dentistry— J. F. Austin, D. D. S., 1885-1887; R. W. Clarkson, A. M., D. D. S., 1887-1888; Emos J. Perry, D. D. S., 1888-1889; R. W. Clarkson, A. M., D. D. S., 1889-1891; J. L. Newman, D. D. S.,

*Informed by Dr. F. J. S. Gorgas.
†Informed by Dr. B. Grant Jefferies.

1891-1892; C. C. Whitmore, D. D. S., 1892 ——.
Professors of Dental Prosthesis—Byron D. Palmer,
D. D. S., 1885-1889; N. J. Roberts, D. D. S.,
1889-1891; C. C. Whitmore, D. D. S., 1891-1892;
B. Joseph Roberts, D. D. S., 1892 ——.

The Dental Department of Kentucky University* was organized in 1886 and is located in Louisville, Ky. Deans—A. Wilkes Smith, M. D.,
D. D. S., 1886-1887; J. Lewis Howe, M. D.,
Ph. D., 1887 ——. Professors of Operative Dentistry—A. Wilkes Smith, M. D., D. D. S., 1886-1887; F. Peabody, D. D. S., 1887 ——. Professors of Dental Prosthesis—Charles D. Edwards,
D. D. S., 1886-1891; Ed M. Kettig, M. D.,
D. D. S., 1891-1892; Charles E. Dunn, D. D. S.,
1892 ——.

The American College of Dental Surgery was
chartered in 1886 and is located in Chicago. Deans
—I. Clendenen, M. D., D. D. S., 1886-1892; J. S.
Marshall, M. D., 1892 ——. Professors of Operative Dentistry—Willard E. Hall, D. D. S., 1886-1887; Ira B. Crissman, D. D. S., 1888 ——. Professors of Dental Prosthesis—J. T. Davenport,
D. D. S., 1886-1887; Ira B. Crissman, D. D. S.,
1887-1888; George Thomas, D. D. S., 1888-1890;
Thomas Rix, D. D. S., 1890-1892; B. J. Cigrand,
B. S., D. D. S., 1892 ——.

The Dental Department of the Southern

---

*Informed by Dr. J. L. Howe.

Medical College* was instituted in 1887 and is
located at Atlanta, Ga. Deans—L. D. Carpenter,
D. D. S.; William Crenshaw, D. D. S. Professor
of Operative Dentistry—William Crenshaw, D.-D. S.
Professor of Dental Prosthesis—John S. Thompson,
D. D. S.

The University Dental School† was established
in 1888 and is located in Chicago. Deans—John
S. Marshall, M. D.; Edgar D. Swain, D. D. S.
Professors of Operative Dentistry—Charles Pruyne,
D. D. S.; G. H. Cushing, D. D. S. Professors of
Dental Prosthesis—L. P. Haskell; — Dennis,
D. D. S.; — Ames, D. D. S.

The Homeopathic Hospital College‡ was insti-
tuted in 1892 and is located at Cleveland, O.
Dean—S. B. Dewey, M. D., D. D. S. Professor
of Operative Dentistry—J. E. Robinson, D. D. S.
Professors of Dental Prosthesis—G. H. Wilson,
D. D. S.; W. S. Jackson, D. D. S.

The Dental School of the University of Minne-
sota§ was established in 1881 and is located in
Minneapolis. Dean—W. Xavier Sudduth, A. M.,
M. D., D. D. S. Professor of Operative Dentistry
—T. E. Weeks, D. D. S. Professor of Dental
Prosthesis—C. M. Bailey, D. M. D.

The Dental Department of the University of

Denver* was instituted in 1888 and is located in Denver, Col. Deans—A. B. Robbins, M. D.; M. A. Bartleson, D. D. S.; P. T. Smith, D. D. S.; Thomas Gaddes, L. D. S., M. D.; George J. Hartung, D. D. S. Professor of Operative Dentistry —John M. Norman, D. D. S. Professors of Dental Prosthesis—M. A. Bartleson, D. D. S.; E. W. Griswald, D. D. S.

The Meharry Dental School† was chartered in 1886 and is located in Nashville, Tenn. Dean— G. W. Hubbard, M. D. Professors of Operative Dentistry—J. P. Bailey, D. D. S.; J. R. Porter, A. B., D. D. S.; H. L. Smith, D. D. S.; S. J. Watkins, D. D. S.; J. B. Singleton, D. D. S. Professors' of Dental Prosthesis—A. P. Johnson, D. D. S.; B. N. Du Pre, D. D. S.; O. G. Mingledorff, A. M., D. D. S.

The Dental Department of the University of Tennessee‡ was organized in 1889 and is located at Nashville, Tenn. Deans—Robert Russell, D. D. S.; J. T. Crawford, M. D., D. D. S.; R. B. Lees, M. D., D. D. S. Professors of Operative Dentistry—Robert Russell, D. D. S.; J. Y. Crawford, M. D., D. D. S.; R. B. Lees, M. D., D. D. S. Professors of Dental Prosthesis—J. Y. Crawford, M. D., D. D. S.; R. B. Lees, M. D., D. D.; J. P. Gray, M. D., D. D. S.

---

*Informed by Dr. G. J. Hartung.
†Informed by Dr. G. W. Hubbard.
‡Informed by Dr. R. B. Lees.

The United States Dental College* was char-
tered in 1891 and is located in Chicago. Dean—
J. J. Angear, A. M., M. D. Professors of Oper-
ative Dentistry—J. D. Robertson, D. D. S.; Dr.
Lazur, D. D. S. Professor of Dental Prosthesis
—G. A. Thomas, D. D. S.

The Dental Department of the Western Reserve
University† was chartered in 1892 and is located
in Cleveland, O. Dean—W. H. Whistlar, M. D.,
D. D. S. Professor of Operative Dentistry—
Charles R. Butler, M. D., D. D. S. Professor of
Dental Prosthesis—George H. Wilson, D. D. S.

The Alabama College of Dental Surgery‡ was
chartered in 1892 and is located in Bridgeport,
Ala. Dean—J. S. Hill, M. D., D. D. S. Profes-
sors of Operative Dentistry—T. M. Allen, D. D. S.
Professor of Dental Prosthesis—F. L. Adams,
D. D. S.

The Detroit Dental College§ was established in
1892 and is located in Detroit, Mich. Dean—
Theodore A. McGraw, M. D. Professors of Oper-
ative Dentistry—George S. Field, D. D. S.; E. W.
Clawson, D. D. S. Professor of Dental Prosthesis
—Dr. F. L. Shattuck.

The Dental School of the University of Buffalo‖

*Informed by Dr. W. H. Prittie.
†Informed by Dr. W. H. Whistlar.
‡Informed by D. J. S. Hill.
§Informed by Dr. H. O. Walker.
‖Informed by Dr. W. C. Barrett.

was instituted in 1893 and is located in Buffalo,
N. Y.   Dean—W. C. Barrett, M. D., D. D. S.
Professor of Operative Dentistry—F. E. Howard,
D. D. S.   Professor of Dental Prosthesis—C. A.
Allen, D. D. S.

The following dental colleges failed to give information: Kansas City Dental College, Columbian University, Dental Department; Alabama
College of Dental Surgery, and New York Dental
School.

DENTAL EDUCATION.

In 1800 there were but one hundred dentists in
the United States, ten years later there were three
hundred representatives of the profession, while
to-day we have a congress of twenty thousand.

What is true of the increase in colleges and
professional practitioners is of necessity true of the
journals and literature pertaining to this great and
wonderful art and science.

Prior to the founding of the Baltimore College
of Dental Surgery (1839) few American dental
practitioners had devoted their attention to dental
literature, but the following dentists gave their
efforts and energies toward the creation of a
nucleus from which since so much has developed:
"Practical Observation on the Teeth," R. Woofendale, 1783; "A Treatise on the Human Teeth,"
Skinner, 1801; "A Treatise on Dentistry," Long-
botham, 1802; "A Treatise on the Management of

the Teeth," James, 1814; "A Practical Guide to
the Management of the Teeth," Parmly (L. S.),
1819; "Lectures on the Natural History of the
Teeth," Parmly (L. S.), 1821; "An Essay on the
Disorders of the Teeth," Parmly (E.), 1822; "The
Family Dentist," Flagg (J. F.), 1822; "Treatise
on the Structures and Diseases of the Teeth,"
Gidney, 1824; "Principles of Dental Surgery,"
Koecker, 1826; "A Physiological Inquiry Into the
Structure, Organization and Nourishment of the
Human Teeth," Trenor, 1828; "Observations on
Neuralgia with Cases," Trenor, 1828; "Observa-
tions on the Importance of Teeth," Fitch, 1828;
"A System of Dental Surgery," Fitch, 1829; "An
Essay on Artificial Teeth," Koecker, 1832; "Re-
marks on the Importance of Teeth," Chewning,
1833; "An Essay on the Disorders of the Jaws,"
Koecker, 1834; "The Family Dentist," Bostwick,
1835; "An Inaugural Dissertation on the Physi-
ology and Diseases of the Teeth," Spooner, 1835;
"Dentalogia," Brown (Solyman), 1835; "Observations on the General Importance of the Teeth,"
Plough, 1836; "Guide to Sound Teeth," Spooner,
1836; "A Public Treatise Upon the Preservation
of the Teeth," Overfield, 1838; "Dental Hygia,"
Brown (Solyman), 1883; "Observations on the
Structure, Physiology, Anatomy and Diseases of
the Teeth," Burdell, 1838; "An Essay on the
Manufacture of Mineral, Porcelain, or Incorruptible

Teeth," Spooner, 1838; "The Dental Art," Harris (C. A.), 1839.

DENTAL PERIODICALS.

"The American Journal of Dental Science" was the first in this or any other country, and was issued on June 1, 1839. The following from a subsequent volume of this journal fairly portrays what the founders of the journal expected to meet and overcome and reads:[*]

"The circular of the publishing committee, E. Parmly, E. Baker and Solyman Brown, sent it forth with many apparent misgivings as to the success of the experiment, and appealed in strong terms to the more intelligent members of the profession to come forward to its support. * * * The Journal was to consist of forty-eight pages, twenty-four of which were to be devoted to the republication of standard works on dental theory and practice. It was to be issued monthly. The need of such a publication was evinced by the promptness with which this effort was encouraged. In the fourth number a list of subscribers, embracing the most eminent names in the profession, was published, showing that there were at that time one hundred and seventy-four subscribers taking five hundred and eleven copies. This may seem a small number, but when it is recollected that this

---

[*]Dental and Oral Science—Dexter, p. 229.

was years ago, when the number of intelligent dental surgeons was very small, it must be regarded as evidence of a remarkably general interest in the undertaking. During the first year of its publication the Journal was conducted under the editorial management of E. Parmly, of New York, and C. A. Harris, of Baltimore, in which latter city it was printed. It was issued with some irregularity at the subscription price of three dollars per annum. At the close of the year it came into the possession of the American Society of Dental Surgeons, which at that time was organized. The title was then changed to that of the 'American Journal and Library of Dental Science.' It was now issued in quarterly numbers, and the subscription price was increased to five dollars. It was placed by the society in the charge of C. A. Harris, of Baltimore, and Solyman Brown, of New York. The editors were to be assisted in their labors by twenty collaborators, whose duty it was to furnish matter for the work and to aid its circulation. From that period until August, 1850, the Journal continued to be issued under the auspices of the society, under the charge of several editors appointed yearly. At the annual meeting, in the year, it was transferred to Dr. C. A. Harris, of Baltimore, the society relinquishing all control of the Journal, which is now a private enterprise."

The following is a list of the various dental

journals which have been issued in the United States and Canada since 1839:

1839 ——. American Journal of Dental Science, Baltimore. Editors: Parmly (E.), Baker, Brown, Harris (C. A.), Piggott, Gorgas (F. J. S.).

1843.* Dental Visitor, Northampton, Mass. Editor, J. W. Smith.

1843-1844. The Dental Mirror, Northampton, Mass. Editors, J. P. and W. R. Holmes.

1844-1857. Stockton's Dental Intelligencer, Philadelphia. Editor, Stockton.

1845.* The Dental Mirror and Brooklyn Annual Visitor, Brooklyn, N. Y. Editor, M. K. Bridges.

1846-1856. The New York Dental Recorder, New York City. Editor, C. C. Allen.

1847 ——. The Dental Register, Cincinnati. Editors, J. Taft, Taylor.

1847-1859. The Dental News Letter, Philadelphia. Editors, J. D. White, J. R. McCurdy.

1847-1848. Dental Intelligencer, Philadelphia. Editor, E. Stockton.

1849.* The Dental Messenger and Lancaster Annual Visitor, Lancaster. Editor's name not mentioned.

1850 ——. The American Journal of Dental Science, Baltimore. Second series.

*Could not determine date of discontinuance.

1851-1852.   Dental Times and Advertiser, Baltimore.   Editor, A. A. Blandy.

1851.* Merritt's Dental Messenger, Griffin, Ga. Editor, F. Y. Clark.

1853-1854.   The Semi-Annual Dental Expositor, Baltimore.   Editor, Solyman Brown.

1852-1853.   The Practical Dentist, Portsmouth, N. H.   Editor, F. Fuller.

1853.* The Family Dental Journal, Albany, N. Y.   Editor, D. C. Estes.

1855-1856.   The Forcep, New York City. Editor's name not mentioned.

1855-1856.   The Dental Observator, New Orleans.   Editor, J. S. Clark.

1855.* The Dental Monitor and Quarterly Miscellany, New York.   Editor's name not given.

1857-1858.   The American Dental Review, St. Louis.   Editor, A. M. Leslie.

1856-1859.   The Dental Reporter, Cincinnati. Editor, J. T. Toland.

1855-1860.   The Dental Enterprise, Baltimore. Editor, H. Snowden.

1858-1864.   New York Dental Journal, New York.   Editors, F. H. Norton and G. H. Perine.

1858-1860.   The Cincinnati Dental Lamp, Cincinnati.   Editor, J. M. Brown.

1859.* The Family Dentist, New York.   Editor, B. F. Smith.

*Could not determine date of discontinuance.

1859 ———. The Dental Cosmos, Philadelphia. Editors: J. W. White, J. D. White, J. H. McQuillen, G. J. Ziegler and Edward C. Kirk.

1860-1861. The Southern Dental Examiner, Atlanta, Ga. Editors, J. P. Brown and G. P. Fouke.

1860-1862. The Vulcanite, New York. Editor, B. W. Franklin.

1860-1861. The Dental Instructor, New York. Editor, E. A. L. Roberts.

1862-1867. The Dental Quarterly, Philadelphia. Editors: A. Tees, F. N. Johnson, G. R. Weden. Continued as Dental Office and Laboratory.

1863-1864. The People's Dental Journal, Chicago. Editors, W. W. Allport and T. T. Creighton.

1863-1873. The Dental Times, Philadelphia. Editors: G. T. Barker, E. Wildman, J. Tyson.

1865-1865. The Dental Circular and Examiner, Albany, N. Y. Editor, B. Wood.

1867 ———. The American Journal of Dental Science, Baltimore. Third series.

1868.* The St. Louis Dental Journal, St. Louis. Editor, J. Payne.

1868-1872. The Dental Office and Laboratory, Philadelphia. Editors, Johnson & Lund. New series in 1877.

---

*Could not determine date of discontinuance.

1869-1869.   The Vicksburg New Era, Vicksburg, Miss.   Editor, W. S. Young.

1869.* The Missouri Dental Journal, St. Louis. Editors: H. Judd, H. S. Case, C. W. Spalding, R. S. Pearson.   Resuscitated as the Archives of Dentistry.

1869-1870.   Houghton's Dental Annual, Poughkeepsie, N. Y.   Editor, C. L. Houghton.

1869 ——.   The Dental Advertiser, Buffalo, N. Y.   Editors, T. G. Lewis, W. C. Barrett.

1872-1872.   The Dental Mirror, St. Louis. Edited by committee from St. Louis Dental Society.

1868-1877.   The Canadian Journal of Dental Science, Montreal.   Editor, W. G. Beers.

1874-1877.   The Pennsylvania Journal of Dental Science, Lancaster, Penn.   Editor, S. Welchens.

1874-1881.   Johnston's Dental Miscellany, New York.   Editors, Johnston Bros.

1877-1878.   The St. Louis Dental Quarterly, St. Louis.   Editors, C. W. Spalding and H. S. Case.

1877 ——.   Dental Office and Labratory, Philadelphia.   Editors, Johnson and Lund.

1878-1878.   The Dental News, Knightstown, Ind.   Editor, T. P. Wagoner.

1878-1878.   The Dental and Oral Science Magazine, New York.   Editor, R. S. Williams.

---

*Could not determine date of discontinuance.

1878 ——.   Items of Interest, Philadelphia. Editor, T. B. Welch.

1879 ——.   The Dental Summary, Macon, Ga.   Editor, J. P. and W. R. Holmes.

1880-1891.   The Independent Practitioner, New York.   Editors: W. C. Barrett, H. T. Byrd, B. M. Wilkerson, G. H. Rohe, G. W. Field, W. C. Barrett.   Continued as the International Dental Journal.

1880-1884.   The Dental Jairus, Sacramento, Cal.   Editor, W. O. Thailkill.

1881-1884.   The Pacific Dental Journal, Sacramento, Cal.   Editor, W. O. Thrailkill.

1880 ——.   The Dental Headlight, Nashville, Tenn.   Editor's name not mentioned.

1881-1885.   The Herald of Dentistry, Brooklyn.   Editor, T. O. Oliver.

1881 ——.   Ohio State Journal of Dental Science, Toledo, O.   Editors, G. Watt and L. P. Bethel.

1882 ——.   The Southern Dental Journal, Atlanta, Ga.   Editors, B. H. Catching and Holliday.

1882-1883.   The Dentist's Beacon Light, La Crosse, Wis.   Editor, Edgar Palmer.

1882-1884.   New England Journal of Dentistry, Springfield, Mass.   Editor, Charles Mayr.

1882 ——.   The Dental Record, Baltimore. Editor's name not given.   Since discontinued.

1882 ——.   Health and Home, Toledo.   Editor, J. Munson.

1883 ——.   The Dental Practitioner, Philadeldelphia.   Editors, C. E. Pike and L. A. Faught.

1883 ——.   The Texas Dental Journal, Dallas, Texas.   Editor, Newman and Storey.

1883-1890.   The Practical Dentist, Elgin, Ill. Editors, Truesdell and Underwood.

1883.*   Caulk's Dental Annual, Camden, Del. Editor, Caulk.

1884-1886.   The Dental Student, Warren, Ind.   Editor, C. A. Rigdon.

1884-1885.   The Dental Review, Charlotte, Mich.   Editor, W. G. Ashton.

1884-1891.   The Archives of Dentistry, St. Louis.   Editors: C. W. Spalding, T. Fuller and Eames.

1885-1887.   The Dental Eclectic, Knoxville, Tenn.   Editor, S. S. Willard.

1885-1886.   Facts, Chatanooga, Tenn.   Editor, E. M. Martin.

1885 ——.   Cincinnati Medical and Dental Journal, Cincinnati.   Editor, F. W. Sage.

1888-1890.   The Practical Dentist, Toledo, O. Editors: C. W. Munson, W. E. Blakeney, F. O. Brake.

1889 ——.   The Dominion Dental Journal.

1891 ——.   Pacific Dental Journal.

*Could not determine date of discontinuance.

1887 ———. Dental Office and Labratory, Philadelphia. Editor, Chupein.

1886 ———. Dental Review, Chicago. Editors: A. W. Harlan, C. N. Johnson, L. Ottofy.

1889 ———. International Dental Journal, Philadelphia. Editors: W. X. Sudduth, James Truman, Joseph Head, G. W. Warren.

1892 ———. The Dental Journal, Ann Arbor, Mich. Editor, C. A. Hawley.

1889-1892. American Journal of Oral and Dental Science, Chicago. Editors: I. Clendenen, G. A. Thomas, G. North, G. A. Stevenson, O. P. Bennett.

1890-1891. The Dental Mirror, New York. Editor, R. Ottolengui.

1891 ———. Dental and Surgical Microcosm, Chicago and Pittsburg. Editor, S. J. Hayes.

1892-1893. The Dental World, Chicago. Editors, E. L. Clifford and B. J. Cigrand.

1892-1892. The Dentist Himself, New York. Editor, Kimble.

1892 ———. The Chicago Medico-Dental Bulletin, Chicago. Editors: W. H. Prittie, J. J. M. Angear, G. North, G. Frank Lydston.

The Dental Tribune, Chicago. Editor, Louis Ottofy.*

While many of these had an ephemeral exist-

---

*The author is indebted to the following gentlemen for the completeness of the list of American dental journals: Drs. J. Taft, H. J. McKellops, Allen Lund, James Truman and E. Kirk.

ence, others have lived well and do worthy service in their strict devotion to the profession.

These various journals did eminent service, in my research, and the records of dentistry contained in these monthly educators are the safe archives of the profession, where we may trace the present dental progress back to those traditional ages, when, too, our favorite vocation prospered and attained that semi-state of perfection.

The following from the pen of the eminent Dr. G. V. Black is appended to the subject of dental periodicals, in the hope that it may cheer the editors and as well the readers of the various dental journals:

"We should not condemn journals because some of the articles are of little value. Much of the thought presented in journals is simply placed on trial, and that which my judgment or the individual judgment of the editor might condemn may prove to be valuable. Many of the better things in literature have been condemned at first reading by learned critics, and have afterward been recognized by the world as models of thought and expression. Neither should we drop the reading of a journal because a number or two fails to interest us. The next number may contain a single article that will be worth a dozen years' subscription, besides compensation for much uninteresting reading. Anyone who fails to read the journals will be behind, not only in his thoughts, but also in his practice."

ARTISTS AND SCIENTISTS.

The days of unwieldy instruments and rude operators are now in oblivion, and in their stead we find instruments and appliances complete in every respect, and practitioners who are as mild a class of men as ever played the lute or sung the songs of love.

It was in the spirit of advancement and a love for the beautiful that the dental infant was taken from the barber shop and raised to the high and ennobling position it now occupies. And who has wrought this most desirable change? None else than the ingenious and persevering modern dentist. The dental office is no longer a prison of torture, but on the other hand is a welcome resort for suffering humanity. To accomplish this good end has cost the burning of much midnight oil, and many of the energetic men who have labored vigorously in this most beneficent cause are now slumbering in the cities of the dead.

It is said that "a poet is born, not made," and this old saw is, in a certain sense, applicable to the dentist. In order to advance and be successful in the dental profession the practitioner must have certain definite qualifications and inclinations. And these essentials must be his or her natural bent of character. Among the requisites of a model dentist

the most important are: That peculiar quality
which makes the successful surgeon, coupled with
mechanical ingenuity, dexterity, studiousness, and
last, but far from least, the patience of Job.   Perhaps the reason why there are so many lamentable
failures among the practitioners of dentistry at the
present time is because so many enter the profession with the sole and whole purpose of gathering
the supposed hoards of money which are believed
to be accessible to its votaries.   But how soon are
the plans and anticipations of these empirics frus-
trated when they find they are not adapted to their
chosen work; unfitted for the science, they drop
by the way.   They are allowed admission into the
labyrinthal highway, but soon are lost and bewil-
dered among the "ologies" and "isms" of the profession.   Dentistry is too great a science for the
gross and unskilled to appreciate the aesthetic
beauties of its art.   The modern dentist must be
in the full sense of the words a "facial sculptor,"
for to his tender care and consideration is left the
moulding of many a scowl or smile.   He must
appreciate the lines of beauty in expression and
discern at a glance the changes necessary in the
different physiognomy to make them charming and
inviting, rather than repellent and false.

In dental prosthesis and dental surgery the
sculptoral genius is certainly afforded an opportunity to exercise his art, since it is in these

departments of dentistry that the study of the face
is most essential.   The face is divine territory
which solicits the prosthetic dentist's sincerest consideration.   The face to him is the window to the
brain, the avenue to mind and character.   The
face is the servant of the emotions; it mirrors the
feelings, and gives expression to impulses.   It is
the visible record, the map of the heart proclaim-
ing the character of the individual to all who can
read.   The symbols of character, which are unmistakably in the face, are not occult and secret, but
are open and plain, that even a child may read
and know distinctly the heart of its owner.   Now,
since the distinctive feature of dental prosthesis is
"restoration," you can readily comprehend why the
dentist who is continually sought to restore faces,
must of necessity be thoroughly prepared to restore
the lost features, and rebuild the symbols of individuality and character.   He must have a clear
conception of the outlines of that which is to be
established, and constantly have the imaginary
ideal vividly before him.   In all grand works of
man the ideal was ever the basis of the real.   In
our own modern city take for example the magnificent Auditorium, with all its halls, porticoes,
entrances, pillars, stairways, arches, balconies and
tower, all was designed by the architect in all its
grand proportions and arrangements before the
foundation stone was laid.   The sculptor who

chiseled from the huge misshapen block the almost
living and breathing figure of Abraham Lincoln in
a park by his name saw in the rough stone the
ideal statue.

Dr. Allport once said: "He who has but moderate ideas of symmetry, harmony of expression and
color, is constantly pained by lack of that artistic
selection and arrangement of artificial teeth which
serve to restore to the face the shape and expression left upon it by the Creator, the absence of
which in artificial dentures stamps him who should
be an artist an artisan—a mere mechanic—a libeler
of the soul—a deformer of the humane face divine."
We can only know how thoroughly scientific,
artistic and technical the restoration of the face is
when we hearken to that great lecturer Fuseli, who
says: "If the nose of Apollo be shortened one-
sixteenth of an inch the 'god of physical beauty'
would be destroyed." If this be true, which it
certainly is, it should lead us to be very cautious
as to the duty of our calling, and ever remember
that the perfect restoration of the countenance
with the original power of expression by art, as to
defy detection, is one of the crowning glories of
dental prosthesis.

This branch of dental science is as a general
thing underestimated, inasmuch as those who
have made a specialty of it have failed to bring to
light the many grand principles that underlie this

most deserving specialty. I can do no fairer justice to the subject of prerequisite qualifications of the student of Dental Prosthesis than by quoting the able scholar, Dr. W. W. Allport, who said:

"It is in Dental Prosthesis the dentist has the greater field for the use of art. It is for him to so construct substitutes for the natural teeth that they will harmonize with the works of the Creator that surrounded them, and be so true to nature in size, shape, color and position that they will not produce discord in the facial expression. There is an individuality in everything that God has made. There are no two blades of grass, no two flowers, two faces, two eyes, nor are there any two sets of teeth that are alike. They may be similar in type, but not in detail, and it is this detail that gives the specific individuality by which we are enabled to tell one from the other. Between these details there is a harmony that makes any one part a fit companion of its surroundings. Any important change in any of these details would, to the extent of the change made, alter the individuality of the original. As there are no two things exactly alike in nature, there can be no exact rules by which anything in nature can be imitated. There are, however, rules which may be aids in producing general outlines, but it is the soul and feeling of the artist that works out the details which gives life to the substitute. A mechanic, pure and simple, may construct a set of teeth and make them serviceable to the wearer, inasmuch as they will fit and be strong and useful in mastication, but only he who has the artistic feeling and skill will be able to select his materials and so adapt them in the mouth that they will harmonize with the complexion and anatomy of the face and be true to nature. From infancy to old age there is harmony in contour, as well as in color, and there is change and adaptation of one to the other at every stage of life. The hair that would be becoming to a girl of sixteen, would not be suited to the same person at sixty. Hence nature changes the color of the hair to be in keeping with the face as age advances. The same is true of the teeth; all change and grow old together, and there is beauty in age only as there is harmony. To attempt, therefore, to make the face look younger or more attractive by making any one part of it appear younger than is natural is a great mistake, for the other parts suffer by an inharmonious contrast which always unpleasantly attracts attention.

"In applying this idea to the selection and adaptation of artificial teeth, it will at once be seen how very important it is that he who gives his attention to this branch of industry should not only be a good mechanic, but should possess that art feeling that will enable him to appreciate the importance of physical harmony. If he does not

possess this quality, he will be a mechanical dentist
only.   His work may be useful for mastication,
but the face will be apt to look 'toothy.'   To
produce this appearance the teeth need not of
necessity be too large for the face.   In fact, artificial teeth are usually smaller than were the natural, and yet they give the appearance of which I
have spoken, as it is usually the inharmonious
color, rather than the size of the teeth that is at
fault.   The first as well as the most lasting impression made on the beholder of the individual will
be the teeth, whereas they should be so thoroughly
in keeping with the rest of the face that they will
attract no more attention than any other feature.

"One of the prerequisites to the study and practice of this specialty is a talent for and a knowledge
of art.   The proportion of good artists who could
have made good mechanics is very large, while the
proportion of good mechanics who could have
made good artists is very small.   A person may
have great mechanical ability, but little or no
artistic sense.   There are few dentists who have
any idea of proportion or feeling for color.   This
is why we see so many mouths filled with abom-
inably unnatural looking artificial teeth, and this
condition of things will never be greatly improved
till more attention is given to art in this department of practice.   It would be useless to attempt
to develop this talent in every dental student, for

probably not more than one in twenty-five, or perhaps fifty, could respond to the demand should they be encouraged to follow dental prosthesis as a calling.

"Artistic ability, therefore, should be among the first requisites to the study and in the practice of dental prosthesis.   It would be far better for those who engage in its practice to have acquired a theoretical, as well as a practical, knowledge of the leading ideas of proper proportions, modeling, drawing and harmony of colors, rather than to have studied so much of medicine as is usually taught in dental colleges."

On this same subject Dr. Joseph Richardson, one of the beacon-lights of dental prosthesis, says:

"Among the unnumbered millions of human beings who have peopled the earth since the dawn of time, it may be affirmed that no two have been created with faces exactly alike.   There is the same aggregate of features and a pervading general resemblance of one person to another, but there will be found as infinite a multiplication of distinct shades of facial expression as there are human faces, and each separate shade of expression characteristic of each one, and distinguishing him or her from all others, constitutes facial individuality. Each separate feature—as the eye, the nose, the mouth, the teeth, facial contour, complexion, temperament, etc.—contributes to this individuality, and no one special feature more, perhaps, than the teeth. There are few more repulsive deformities than those inflicted by the loss of these organs, and none more fatal to the habitual and characteristic expression of the individual. It is the special mission, as it is the first and highest duty, of the dentist to preserve this individuality intact, and an equally imperative duty to restore it as perfectly as possible when impaired. To fullfil in the most perfect manner possible this most difficult of all the requirements of prosthetic practice implies an art culture that is competent to interpret the distinct play of features associated with individual physiognomies, to differentiate individual temperaments, and make available the sculptor's and painter's perceptions of the subtile harmonies of form and color. To the failure or inability to properly comprehend the practical import or significance of individual characteristics, so far as the fixed expression in the teeth and the consequent failure to conform our methods of replacement to the imperative requirements of art, may be fairly ascribed the deserved reproach into which prosthetic practice has fallen, and not, as is generally charged, to the employment of any particular material or methods concerned in the mechanical execution of the work."

The grandest thesis on the "art in dentistry" is the following most able article by Dr. Kingsley:*

"That 'Dentistry is a Science and an Art' is a statement that at this day seems hardly necessary to reiterate.

"The phrase has a royal sound, and its frequent repetition shows that it is a favorite expression; and yet among the multitudes who earn their livlihood by practicing dentistry, how few know the full meaning of the words, and how little there is in their practice to justify the assertion.

"To judge by its fruits, how much more there is of empiricism than of science, and how much more of rude and bungling mechanism than of art. Nevertheless, dentistry is a science and an art, and the researches of the past four years alone, together with the contributions to its literature, give a legitimate claim to the first part of the proposition. Months and years have been spent in the prosecution of it as a science and volumes record its results, but as an art, capable of taking rank as one of the fine arts, dentistry has rarely, if ever, found in our journals an advocate—has rarely found else than a flippant consignment to the workshop, where the very idea of art comprehends only ordinary mechanics.

"As a consequence artistic dentistry has never risen, except in rare individual cases, to anything above mechanical dentistry. The very term by which the department is known is often used as one of reproach, and the productions of these dental mechanics are a standing disgrace. In every assemblage, public or private, on the street, in the drawing-room, or wherever we may turn, we see displayed their hideous deformities. It becomes a serious question, whether the art of dentistry, aside from operations upon the natural teeth, has, with all the inventions and 'improvements' of the last decade, advanced one iota.

"The operative department has assumed to be the department, par excellence and per se, and we see the results in the education of the new professional generation, who ignore as unworthy their exalted talents any knowledge of mechanical dentistry, not realizing that a mastery of all its elements will do more to educate and qualify them for perfection, even in the one department, than any other course that could be pursued. We believe that it can be demonstrated beyond a peradventure that the ignored and despised branches of dental science can lay a well-grounded claim to be considered as a fine art, capable of the highest idealization, ranking side by side with poetry, music, painting and sculpture, capable of appealing, though in a more limited manner, to the same sentiments and emotions, and requiring for their expression the identical talent and same imagination which characterize the votaries of her predecessors.

"With the ancient Greeks all works which exhibited skill were called works of art, and to the present day the term art, in its broad signification, is applied to every skillful, physical, or intellectual performance, from the making of a shoe to the modeling of a statue, from the pantomime of the stage to the oratory of the forum. But as the arts have multiplied, terms of distinction have become necessary, as fine arts and mechanical arts with all their subdivisions.

"The distinguishing characteristic of the fine arts is their ideality. In this the line of demarkation between them and the mechanical and all other arts is unmistakably distinct.

"It is for this feature we look in any work that claims this high rank, and by this standard we judge of its pretension. The mechanical arts are distinguished for their physical utility. They may demand consummate skill for their execution; they may require for their development rare inventive faculties, and their combinations of mechanical principles may be truly wonderful, but their individual works require but little effort of the brain in their reproduction; education in skillful manual labor without the capacity to originate a single new idea is all that is required. The laws which govern their reproduction are those of mathematics, and to be able to copy a given form with exactness is the sum of the talent required.

"They may be directly of more practical value

to mankind, but they can make no appeal to the finer emotions of our being. In all that excites the imagination, that calls into action the affections, or leads the mind away from the contemplation of the material and sensual, they are dumb.

"In like manner the feats of the acrobat and juggler excite our wonder and admiration, but like the true mechanical arts, have not an element of ideality in them.

"The ideal or fine arts, therefore, may include poetry, music, painting and sculpture. These require for their development the possession and exercise of the same mental faculties, are governed by the same general rules, and have one common ultimate object.

"Poetry, of all the arts, stands deservedly at the head, because the most subtle and at the same time the most potent in its influence. It is the least material of all, the farthest removed from sensible objects; it has the greatest scope, admitting the treatment of all subjects, and its power over the imagination is the most complete. In regard to the objects of the visible world, one can not conceive a greater distance between what it depicts and the manner of depicting. By the combinations of words used alone, used as language, poetry gives full expression to every idea, from the most powerful to the most delicate, and presents scenes so vividly as well as so variedly to the mind of the reader that the impressions are both pleasing and permanent.

"Music takes precedence over painting and sculpture in the order of the fine arts, its claim to superiority being based upon inherent qualities of humane nature, by whose control all that appeals to our emotions is more regarded than what appeals to our understanding. Objects of sight, while they please the eye by their color, form an arrangement, affect mainly the mind, and only indirectly approach the heart.

"But music speaks directly to the soul. Its notes are the reproduction and the refinement of natural tones of pleasure, anger, fear, distress, etc. These tones music cultivates, and by prolonging and combining them expresses feelings and awakens every variety of emotion into sympathetic activity. Thus the tender pipings of the flute call forth the love of enjoyment; the clear bugle notes find responsive echoes in the hearts of mountaineers and huntsmen; the orchestral overtures send thrills and throes through fancy-loving souls, and the choruses of Mendlesohn and Mozart fill man's spirit with ecstasy of joy or wonder, or with solemn awe.

"More than any other art is music universal. A language without words, music addresses the feelings by tone, everywhere understood and unmistakable; and it leads forth the emotions to an enjoyment which words can not give nor can even express.

"Painting, in the order of the ideal arts, holds the third rank. It is more limited in its scope than either of its predecessors, and expresses its sentiments or tells its story by the color and lineal appearance of bodies. But while it can not undertake the relation of a succession of events in one representation, it has the advantage over poetry, in that its language is universal, easily recognized, and needs no interpreter or translator of its meaning.

" 'A true picture its own story,' and in no way does a common mind receive more vivid and lasting impressions of portrayed events than by the production of this art.

"Many pictures there are whose sentiment is sacrificed to that which merely pleases the eye, the gratification of the senses by the harmony of forms and color being the highest aim of the artist. It is this degeneracy into mere physical representation and the limited nature of its power that places it as an ideal art in the rank that it occupies.

"The art of sculpture, while requiring for its execution the very highest order of mental faculties, is generally placed in the scale of ideal arts, below that of painting, because, represented solely by form and without the aid of color, it is more

limited in its subjects and more material in its
exhibition.

"In the selection of the human figure, the most
perfect of all forms, it finds its grandest achievements in depicting all gradations of intelligence,
affection, sentiment, action, or passion, sublime,
heroic, or tender, and in all orders of beings, from
the exhalted supernatural to the lower gradations
bordering on the brutes. Although the most
limited in its scope, it is not so liable to degeneracy as its sister art of painting. To be successful its delineations must be above that of simply
copying nature, and its power over the beholder is
often greater than any other art could give to the
same subject.

"It is the most enduring of all arts, the material
chosen for its medium being the most independent
of all the mutations of time. Of the peoples of
the by-gone ages, the only records left to us that
give even a passing glimpse of their existence are
their sculptured monuments.

"Like painting and music, its language is universal, and like painting, when it ceases to appeal
to the imagination and seeks only to please the
senses by the beauty of form, it degrades its character and fails of its true mission.

" 'Nothing brings people of other nations so
vividly before us as their works of art. They tell
us of their religion, of their social dwellings and

customs, of their advance in civilization and religious culture. To the study of ancient history a knowledge of the arts as practiced by different nations is indispensible. Language is more or less subject to change and decay, and the significance of many expressions is lost to those who do not use it as a vernacular tongue. Many nations have existed with no written language, but the Almighty in blotting them out from the face of the earth has permitted their works of art to live, from which something may be learned of their rise, progress and character.

"Architecture is very commonly regarded as one of the fine arts, and ranked next to sculpture and painting. Modern architecture is addressed to the eye and the intellect alone, and not to the imagination, and the ideal character necessary to distinguish a fine from a mechanical art is wholly wanting. Wherever this element does exist in buildings of the present day it has borrowed it from sculpture. But architecture among the ancients was one of the earliest of symbolic languages. The pattern shown to Moses on the mount by God, as the model for the building of the tabernacle, embodied the very highest order of ideal art.

"Every post, every board and every bar, every ring and every curtain, were typical of man's redemption from sin, and spoke a language unmistakable and clearly comprehended by the Jews; and so with every monument of the ancient heathen connected with their religion, every temple, every idol, and every altar appealed to the imagination and spoke an ideal language.

"Architecture was then emphatically a 'fine art,' but at the present day it is but an imitation of the dead past. The powers that called it into existence are gone, and the emotions to which it gave birth have died out. We imitate its corporeal form, but the spirit that gave it life is forever departed.

"We have been thus specific in our description of the fine arts for a more thorough understanding of grounds upon which we shall base the claim of dentistry to be ranked with them. No performance of the dentist can make any pretension to be a fine art, separate and distinct from all others, but as a subdivision or speciality of one of the arts dentistry is entitled to a consideration which it has never received. We shall endeavor to show this alliance, and prove that, so far as its scope will allow, it is governed by the same general rules which control its allied arts.

"Dental practice, by an inherent law and by common consent, is divided in the main into two departments, the one commonly termed the 'Operative,' which is made to include all efforts for the preservation of the natural teeth and all surgical operations in the buccal cavity; the other, called 'Mechanical,' which includes the making of all appliances for the correction of deformities of the buccal cavity, but practically the making and inserting of artificial teeth.

"In the practice of operative dentistry, as has been before intimated, there has grown up an unwarrantable assumption that all that was refined and cultivated, all that was worthy the exercise of our noblest faculties in the pursuit of our profession, was to be found in this department, and mere mechanics, wholly unqualified by education in science and art, were deemed capable of practicing the other. The only performance of the operative dentist which requires a talent and skill equal to the mechanical arts is the introduction of fillings into the cavities of decay, and this skill is mere manual dexterity guided by good judgment. Its highest achievements at the present day are in the so-called contour fillings made of gold, in which an attempt is made to restore the form of a tooth injured by accident or decay.

"Contour fillings, when carried to its highest state of perfection in restoring the actual or the typical form of the lost organ, can present no stronger argument to be considered an artistic performance than that of a copy or an imitation. If a copy it is purely mechanical; if an imitation of the typical it may lay a faint claim to ideality.

This is true when carried to its ultimate; but practically nine-tenths of what are called contour fillings
are not entitled to any such distinction.   Nuggets
of gold they are, built on to deformed teeth, carried
in many instances far beyond the borders of decay;
lapping over and building upon sold enamel to a
general level, obliterating all inequalities and all
character, and failing most completely to illustrate
the possession of any other talent than the skillful
manipulation of gold; excellent advertisements of
the craft they undoubtedly are, but are certainly
of very questionable taste.

"Every tooth has an individual character and
expression, not only in harmony with every other
in the same mouth, but by the same divine law, in
harmony with the features and the character of the
creature, be he animal or man.

"The physical characteristics are so marked
and prominent that the merest novice has no difficulty as a rule in locating any human tooth that
has been removed from its fellows, and yet of the
attempts at restoration of any large portion of the
crown of teeth by dentists how few there are that
bear any very close resemblance to the original
form of the lost part.

"If a cast were taken of the restoration and
examined separately, how few would identify it as
being any portion of any tooth.   The cusps, the
depressions, the sutures, the easy and graceful out-

lines, and all that marks the individual tooth are
wanting.   *   *   *   *   If sculpture necessarily
ranks below painting in the scale of fine arts because more limited in its range, and painting for
the same reason below poetry, we must therefore
place all operations on the natural teeth, as artistic
performances in rank, below that of the substitution of artificial ones.

"As an art it is but a department of sculpture.
Form in individual members, form in grouping and
arrangement, and form as a medium of expression
are equally the distinguishing characteristics of
both sculpture and dentistry."

Thus a good dentist should, indeed, be a man
of great refinement, of artistic conception, with a
true sense of the proportion of things and of the
harmony of colors.   We have only to look at the
teeth people often wear to notice that this is not
very often the case.   It must be remembered that
in nature there is a great beauty in the irregularities, in what is often called the ugliness of shape
and color.   Because an even row of very white
teeth is the ideal, it does not prove that such teeth
suit everybody.   What can be more ghastly than
an old, decrepit personage, with a bad complexion,
who wears a double row of splendid white teeth?
What is more rediculous than one white, spotless
artificial tooth standing in the midst of yellow and
partially decayed real teeth?   Or, again, what a

lobsided effect is produced if natural teeth on one side of the mouth grew irregularly, while on the other side artificial teeth have been fixed up in regimental order. Yet how few people are there who, having artificial teeth, have the good sense to ask that these teeth should be just as imperfect in shape, position and color as the real teeth were they are destined to replace?

If we have not ideal teeth the probabilities are that there are many other things in feature and complexion which also are far from being ideal. And the introduction of one or more ideal teeth where the surroundings are anything but ideal is no improvement. It creates a discordant note, destroys the harmony which prevails even in ugliness, and renders that ugliness more evident and more unpleasant. But it requires a high conception of true art to thoroughly appreciate these principles and apply them successfully in practice. It is, therefore, not surprising to find that distinguished dentists are the constant and appreciated friends of men of art and letters.

Nothing but careful study and experience can develop the artistic sense to a degree that will enable one to forecast the shades of expression that it may be desirable to give the patient by well-considered alteration in the length and edge-shape of dental organs.

In short, he must be, as Dr. Marshall says,

"thoroughly conversant with physics, with mechanics and with metallurgy. He must acquire a delicacy of touch and a manipulative skill of the very highest order; his eye must be trained to a keen perception of form, color and harmony, and his hand to execute the thoughts of his brain—in other words he must be an artisan, artist and physician all in one."

DENTISTRY AS AN ART AND SCIENCE.

These two words seem to be used by the large majority of writers and speakers regardless of true diction. Authors often use the former word when the meaning of the latter is inferred and vice versa. There is an art of dentistry and a science of dentistry. This confusion does not only exist in the dental literature, but is also present in other departments of scientific research, and is especially found in German works of art and science. With the German writers the etymology of the word for art secured a long continuance for this ambiguity. The word "Kunst" was employed indiscriminately in both the senses of the primitive "Ich kann," to signify "I can," and later the word "Wissenschaft" came to use, and designated "Ich kenn," or "I know." But the word "Wissenschaft," or science, was not coined until late in the 17th century, and it is largely due to the recent use of this word that the loose interchange exists between it and the word kunst or art.   On this interesting subject the following:*

"The term dentistry is appropriately used to denote the business or the place of business.   As a business it includes both the science and the art of treating, preserving, and artificially substituting teeth.   Dentistry is both a science and an art, while it evolves much that is purely mechanical. Science is the architect, mechanical art the builder; science discovers the want and the means of supply, art attains the ends.   Science prevails, art avails.   Science, like 'charity, seeketh not her own, vaunteth not herself;' but art is fond of pecuniary reward.   Hence the scientific are few, the artists many.   Consulting ease of enunciation, we say 'arts and sciences'; but in fact science occupies the foreground; art following reaps the harvest.   Science acquaints herself with the whole economy of animal teeth, with their matter and their manner, their origin and their end, with all their facts, in all their forms and in all their varieties, their connection with and adaptation to the varied modes of animal existence; and why properly formed teeth are unnecessary to some forms of existence and indispensable to others; why wanting in reptiles, creeping things, and flying fowl; why the higher order of animals scarcely subsist without them; why, with many, the duration of the teeth is the measure of their existence.

"Science would know why the infinite variety of form and size, from the thistly jaw of the smaller aquatics, to the movable envenomed fang of the serpent, up to the powerful war weapon of the tiger, the lion, the bear, the whale, and the mastodon; and why the peculiarity in number, in structure, and form of the human teeth; why both deciduous and permanent; why come in the time and order they do; why partly vital and partly not; why incapable of extension and growth and of repairing their fractures or abrasions; why subject to decomposition and rottenness; why the medium of so much pain; and what are remedies for their diseases? Science, too, must understand the chemical as well as vital organism of the teeth; the matter and the manner of their composition and formation, the arteries, veins and nerves, and all the ramifying capillaries and nerve-fibrils, as well as the corpuscles, tissues, and granules that enter into their composition and organism. Science would understand the entire functions of the teeth and the means of securing them in their normal condition and appropriate use.

"Mechanical dentistry, as such, has little or nothing to do with the animal economy; but as an art dentistry seeks the hand of science, and by her would be led and guided. It would supply the artificial remedies and mechanical appliances agreeably to the teachings of science, adopt and adjust all her fixtures and mechanical powers and agents, remove all obstructions to the health of teeth, and supply losses in accordance with the laws of nature as discovered by science.

"The dentistry of the operating chair is no less mechanical, though more artistic, than that of the laboratory. Indeed, mechanical skill and artistic display make their happiest efforts in perfecting the form and external condition of the natural teeth. While some operations are more strictly surgical, the whole class of operations of the dentist, when performed in accordance with scientific principles, deserve, as they have received, the appellation of dental surgery."

"There is a science* of medicine and there is an art; and it is possible for a scholar to be deeply interested and highly cultivated in the grand science without pretending to be even a tyro in the art. He may have spent years in his labratory in the investigation of the supposed origin of disease without having the slightest skill in detecting or distinguishing the simplest malady at the bedside. He may, in short, be a deep student of the science without being an expert in the art. The science and the art are to this extent divided.

---

*C. M. Wright, D. D. S., Ohio State Journal of Dental Science vol. III., p. 17.

A physician might, in his leisure hours, acquire an intimate knowledge of the science of dentistry; he might become expert in making preparations of dental tissues for histological studies, for the investigation of the theories in regard to dental caries; he might become well versed in a knowledge of the general pathological conditions of the teeth, and yet, if called to the dentist's operating chair, could not diagnosticate an incipient alveolar abscess, nor an inflamed pulp, nor a pulp chamber filled with gas; not to mention the excavating of a carious cavity and replacing with any dentist's materials. He does not possess the Art of Dentistry.

"The science of dentistry is a part of the science of medicine. I wish to say this distinctly and dogmatically. It is a fact not affected by the disputes of doctors or dentists. The science of dentistry is a branch of the science of medicine in its broadest signification; it springs from the same root; it is nourished from the same fluid that circulates in the parent trunk and all the other branches. I refer only to the science, so-called, and not at all to the art. Here we must make a distinction. The art of medicine and the art of dentistry have but very little in common as generally practiced in this country. And the art itself is so distinct from the science in dentistry and medicine that men do become skillful and successful doctors and dentists with but a limited capital of general science—but, in its stead, cultivated perceptions and trained fingers. The young graduate of several native and foreign schools and universities may really possess very much more knowledge of what is called the science of medicine than the old practitioner, and yet you and I would prefer to trust the man of art rather than the man of science. The old practitioner, if possessed of natural talent (and this, however important, is like beauty—possessed only by a few, though desired by the many), has his senses cultivated to the point of intuition. He sees at a glance—he arrives at a correct conclusion instantaneously—by an unknown·and unregarded mental process, that only experienced observations, practiced judgment, cultivation and habit can give. The art is the important part. The science has no other object, finally, than the practice of the art."

Dr. Bonwill even goes further into the analysis of the etymology of the use of art or science in connection with dentistry and says:*

"Every year there is fresh vennom hurled against it,† or any reference thereto, and it would seem to have fallen from the high eminence which it had in the early days. The good old name has a very strange synonym (prosthesis), which might

---

*Western Dental Journal, vol. II., p. 11.
†Mechanical Dentistry.

be well if it were confined to the supply alone of artificial dentures. A moment's reflection, however, will show that when we attempt to divide dentistry into operative and prosthetic or mechanical, we do injustice to our profession, and we assume a false position for our great art, which is founded almost exclusively on our abililty as mechanicians, and without which, to a great degree, we would be helpless.

"What is science?

"Science is a systematic and orderly arrangement of knowledge.

"Facts and truths are the ultimate principles.

"Pure science is a knowledge of forms, causes, or laws.

"While both science and art are synonymous, as they investigate truth, science is restricted to the inquiry for knowledge and art for the sake of production.

"The most perfect state of science will be the high and accurate inquiry, and the perfection of art the most apt and efficient system of rules, art always throwing itself into the form of rules.

"What are the sciences which have been so long recognized as seven in number?

"Grammar, logic, rhetoric, arithmetic, geometry, astronomy, and music.

"You perceive that medicine, law, and theology are not mentioned. They are professions."

MECHANICAL DENTISTRY VIS. DENTAL PROSTHESIS.

In 1883 the American Dental Association appointed a committee on dental literature and nomenclature, and the committee's report in the following year read in part as follows:

"In the art and science of the dental profession there are many distinctive names and phrases. For the most part, perhaps, these are well chosen, and as expressive as is possible in the present state of our language. There are, however, quite a number of nominations of very questionable propriety. To a few of these we now propose to call attention.

"MECHANICAL DENTISTRY.—This phrase has long been used to designate that part of dental science and art which has to do with the manufacture and insertion of artificial teeth and all that pertains to it. So far as the needs of mechanical skill are concerned this name may be sufficiently distinctive, but it is not sufficiently extended in its reach. Some may execute well in a mechanical aspect, and yet signally fail in the production and application of substitutes for lost teeth and the restoration of adjacent parts. The name that will embrace both these ideas, namely, that of mechanism and of art, that shall enable the application of it to the parts here indicated, is that which should be employed, and in a search of some extent no better term occurs for this name than 'dental pros-

thesis.' Some have been disposed to change the
form of the phrase 'prosthetic dentistry.' The
other is a better phrase, we think. This nomi-
nation is slowly but surely making its way into the
literature of our profession. It will probably be
used ere long to the exclusion of the phrase 'mechanical dentistry.'

"OPERATIVE DENTISTRY.—This name or phrase
is usually employed and properly so to designate
operations upon the natural teeth for their preservation, or rescuing them from the ravages of
disease. It embraces operations upon and treatment of the natural teeth, and is sufficiently clear
and distinctive. There has been a little disposition in some quarters to drop this name and embrace everything done in the mouth upon both the
hard and soft tissues in the name of 'oral surgery.'
This may with great propriety be applied to all
operations upon the soft and hard parts in and immediately about the mouth, but the nature of the
operations upon the teeth is so different from those
performed upon the other parts of the mouth as to
justly entitle them to a distinctive name. There
is sometimes an effort made to have a name embrace too much, and this would seem to be a case
of that kind."

The committee are to be congratulated for
having partially shown why the term mechanical
dentistry is not properly applied, and their recom-

mendation of the term "dental prosthesis" should
receive a more general regard, especially by the
dental colleges and dental authors.   But the report
is not without grave mistakes, and the primary
one looms up in the term "operative dentistry."
On close investigation I fail to learn of the real
difference between the terms dental prosthesis and
operative dentistry, as suggested by the committee.
Does the committee infer that the dentist practicing dental prosthesis does not operate? and also,
that the doctor who makes a specialty of operative
dentistry does not practice and follow the principles of prosthetics?   In both of these grand
divisions of dentistry the dentist operates.   For
example, in prosthesis, if when the dentist sets a
Richmond or other crown does he not operate;
does he not prepare the root of the tooth; does he
not operate on organic matter?   And, on the contrary, when the so-called operative dentist inserts
a large gold filling, does he not "add to," "replace," "affix," or "restore?"   If so, he is purely
practicing dental prosthesis.   The terms dental
prosthesis, as now recommended by this national
committee, is correct, but we are obliged to have
a future committee which will coin a proper term
for the now so-called operative dentist.   The
following from the Dental Cosmos is interesting on
this confusion of dental nomenclature and reads:

"We have divided dental art into medical,

surgical and prosthetic. The two first connect it
with a healing art and demand a medical education;
but the characteristic element of dentistry is its
prosthetics—just as therapeutics is the specific
function of the physician. To remove diseased
structure and replace it with gold—to remove
diseased organs and replace them with porcelain—
is the work which demands nine-tenths of the dentist's time; success in which gives him his reputation.

"You may call the one operative dentistry and
the other mechanical dentistry, if you choose; but
each consists in a series of operations and both are
purely mechanical manipulations of material, by
means of instruments; both, also, are acts of replacement. I think it, therefore, more exact and
descriptive to subdivide the peculiar work of dentistry into—structural and organic prosthesis.

"Both are so difficult that highest excellence in
either department is rare, and scarcely ever do we
meet with a double 'first-class.' Hence the practice of dentistry is itself subdivided, following the
example of its parent art. But subdivision does
not imply less honor in the pursuit, so long as we
recognize, in preparing for it, the necessity of a
knowledge of the whole art of which it forms a
part."*

That the term mechanical is and should be

*Dental Cosmos, vol XVI., p. 500.

obsolete, as used to designate dental prosthesis, this clipping from an able treatise by the famous Dr. John Allen is of worth:*

"It is generally conceded that America has better dentists, and more of them in proportion to population, than any other nation on the globe; they are doing all in their power to stay the progress of the loss of human teeth with which we are afflicted, and their timely aid has been crowned with unparalled success. But still, the immense number of teeth that are annually lost causes a great demand for artificial dentures—a large and important branch of dental practice.

"In the construction of these substitutes we should approximate as nearly as possible to the natural organs, keeping the mind's eye upon at least three important points to be attained, viz.: mastication, enunciation and restoration of the natural form and expression of the teeth, mouth and face. But how to attain these ends under all the different circumstances we meet with in this department is a problem not so easily solved as many suppose; for artificial dentistry differs widely from any other branch of business pertaining to mechanism.

"For example, the mechanic works by well-known rules and laws that have been long and well established, and he follows the same routine with

*Dental Cosmos, vol. XI., p. 485.

his rule, compass and square that thousands of others did who preceded him, all producing the same practical results. The architect of the present day has the same well-established principles to guide him now that were employed by the ancients.

"The different styles of architecture known as Doric, Ionic and Corinthian were the favorite orders among the Greeks and Romans in their most palmy days, and these orders, with slight modifications, have been transmitted with mathematical precision to the present time.

"Watchmaking is all done by fixed rules which the workmen have only to follow in order to produce good time-pieces. Thousands of those little wheels are made just alike, and placed in cases in precisely the same relative position to each other, and all will serve exactly the purposes intended. Numerous branches of mechanism are successfully pursued by men of moderate capacity, by simply adhering to certain fixed rules and principles in executing their work.

"But, in the construction of artificial dentures, there are no fixed rules to guide the dentist, for he has no two cases alike; therefore a rule that would apply in one instance would not hold good in another. If he should make a thousand sets of teeth, all just alike, upon one model, he would find but one set out of that whole number that could

be worn, and that only by the one person from
whose mouth the model was taken. Therefore,
instead of working by rule and scribe, as the mechanic does, the skillful dentist is ever devising
ways and means to meet the various requirements
of each particuliar ease.

"The teeth give character to the physiognomy
of persons; therefore as great a variety of expressions should be given them as there are individuals for whom they are intended. Here the
skill of the artist is required in order to avoid an
unnatural contrast, that would lead to detection;
for you will recollect it is the height of art to con-
ceal art.

" 'The dentist who is a true artist is not ambitious to have his work bear the impress of artificial teeth, but on the contrary, that they should
possess that depth of tone, natural form, and truth-
ful expression which characterized the natural
organs.

" 'Varying the position of the teeth will change
the appearance of the mouth, just in proportion as
they differ from the natural teeth. Hence, in
many persons, their former expression is entirely
lost and distortion has taken the place of symmetry.

" 'A want of taste and skill in the construction
and adaptation of artificial teeth results in rude
and graceless work, which contrasts widely with

that of the true artist, who carefully studies the tone, position, and expression of every tooth, and restores the harmony which nature had originally stamped upon the features of his patient.

" 'A few slight touches of the brush in the hands of a skillful artist, will change the whole expression of his picture. So with the teeth; a slight inclination, outward or inward, or variation in length, will change the entire expression of the mouth.'

"Again, the deflection of the various muscles of the face, consequent upon the loss of the natural teeth, presents another class of physiognomical defects, which also comes within the range of dental practice; and the time has come when the dentist is expected to raise the sunken portion of the face to their original contour by artificial means.

"Whether this could be done without injury to the muscles thus raised remained a problem to be solved by an American dentist. This question being settled for all coming time that no injury results from wearing properly constructed dentures with attachments for this purpose, it has now become a practical and important feature in dental prosthesis. The sunken portions of the face can be raised by means of attachments or prominences made upon the denture of such form and size as to meet the requirements of the various cases that are presented to the practitioner.

"In view of the facts here presented, and of what is required of the dentist of the present day we would urge the importance of a higher standard of qualifications in this department than seems to have been attained by a majority of those who are engaged in this branch of our profession.

"The face, as you are aware, is formed of different bones and muscles which give it shape and expression. When the teeth are lost and a consequent absorption of the alveolar processes takes place, several of these muscles are liable to fall in or become sunken, in a greater or less degree, according to temperament. And, in order to restore them to their former position, the dentist should be familiar with the form and position of every bone of the face, and know the origin and insertion of every muscle, what ones to raise, and where to apply attachments to the denture; otherwise he may produce distortion instead of restoration, by underlaying other muscles than those intended to be raised.   Here again the artistic skill of the dentist is brought into requisition. He should study the face of his patient as the artist studies his picture, for he displays his talents not upon canvass, but upon the living features of the face; and of how much more importance is the living picture which reflects even the emotions of the heart than the lifeless form upon canvas? In raising the different muscles of the face the true artist will carefully avoid producing a stiff, restrained, or puffed appearance. He will place the prominences upon the dentures in their proper position, and make them of such form and size as to allow the muscles to rest, move or play upon them with perfect ease, that they may again reflect those sensitive emotions which tell of the inner workings of the mind. Or, to use the language of Shakespeare, 'Your face, my Thane, is as a book where men may read strange matters.'

"Another important consideration in the construction of artificial dentures is, that the materials of which they are formed should be incorrodible or chemically pure.   *   *   *   *   *

"This purity of materials we have in the continuous gum work, when properly made, as none of the materials used are corrodible in the slightest degree in the mouth. Again, all the essential points here referred to can be attained by this mode of constructing artificial dentures. But too much reliance should not be placed upon the mode, for however perfect this may be in itself, artistic taste, skill and judgment are necessary to direct the operator in his manipulations.

"Two artists (so called) may employ the same method, use the same paint, brushes, canvas, etc., in painting a picture. One will produce a perfect prototype of nature, with all the delicate shades and tints peculiar to their art, while the other makes

a mere daub that is worthless. The same difference exists among men in various other branches
of art and science. In conclusion, allow me again
to urge upon our brethren the great importance of
bringing into requisition a much higher order of
talent in the artificial branch of our profession than
has heretofore been employed by a large number
of dentists, whose ambition prompts them to do
the cheapest, not the best work."

this profession is prominent among the many new
occupations opened to women. In keeping with
the enlightened spirit of the age the question of
sex in labor is being lost sight of, in the vast more
important consideration of the quality of labor.   It
is the work per se, not the work per sex, that is
commanding the attention of an educated and dis-
criminating public.   The question naturally arises,
is women fitted for this new field? Is she endowed
by nature with those qualifications necessary for
the labor in the vocations; is she physically and
intellectually adapted, and can she attain proficiency in mechanical skill and mathematical precision? These questions are open to debate, yet,
if we permit the general public to be the jury and
allow the suffering masses, who have received professional care at the woman's hand, to plea, I am
confident that the decision would be a general and
grand triumph for the ladies of our profession.

The following sentiment on this interesting
subject, from the Dental Cosmos, is fully corrobo-
rated by the liberal representatives of our profession:

"The time has long since passed," says the
editor, Dr. Edward Kirk, "when the availability
and fitness of woman for the practice of dentistry
can be successfully questioned, and whatever may
have been the difference of opinion as to her qualifications, both physical and mental, for this work
when measured by the standard of male requirements; the fact remains that in dentistry, as in all
branches of the great healing art, women has found
and successfully occupied a field of usefulness in
which the sum total of those distinctively feminine
qualities, which go to make up an ideal womanhood, have been invaluable, and are, after all, the
essential factors of her success in these departments. The question," continues the editor, "is
not whether she is capable of doing her work from
a man's standpoint and by masculine methods, but,
is there a sphere of usefulness in our profession
which she can pursue and properly fill by virtue of
her womanhood, and achieve success as woman?
The affirmation of this has been demonstrated by
experience and we believe that the ennobling influence of her activities in dentistry will be in-
creasingly felt in the course of time to the whole
dental profession."

In this memorable American year, 1892, the
Woman's Dental Association was organized in
Philadelphia, and like the many similar organiza-
tions, its object is to promote the professional

interests of its members through the advantages
which association confers.

Dr. Truman introduced the lady dentists in a
valedictory delivered March 1, 1866. As a matter of history the following extracts from the valedictory are subjoined:*

"When the professions cease to be objects of
interest to the human intellect, that intellect may
be said to be in its decadence, if it has not already
lapsed into barbarism. The professions lead the
civilizations of the world; as they advance the
nations advance to higher intellectual attainments.
I, therefore, welcome all (who feel they have something to do therein) to the profession to which I
belong, and gladly would I welcome still more than
the world generally concedes have a right to be
there.

"The recognition of the right of every human
being to an equal share in the privilege we enjoy
has not yet become a principle of faith and practice as I think it should. We say to one-half of
the human family stitch, stitch, darn stockings,
make shoes for a shilling, stand behind counters
for two or three dollars a week, do anything, but
don't enter the sacred precinct that we have
marked out for our peculiar benefit. Every human
soul has certain qualities; these should mark its
pathway through life. Talent is of no sex, color

*Items of Interest, vol. XI., p. 530, 1889.

or clime; but is an inheritance from the Creator, given to be fully cultivated in the direction it leads. Hence, in my judgement, any attempt to cripple the aspiration of a God-implanted intelligence is unworthy the age in which we live and is but little short of blaspheming against the Creator himself. As we keep any number of the human race in a condition lower than ourselves, just in that proportion will the degradation be a mill-stone around our necks. The reserve of the proposition is also true; as we advance the masses in intelligence and the means of acquiring information and pecuniary reward for labor, will the civilization of all be increased. Hence, as an individual, I welcome all classes to the profession of which I am a member, and would make but one requirement: Do you believe you are qualified for it and can do better in it than any other position in life?

"Entertaining these views, I rejoice that dentistry, though the youngest of the professions, has welcomed women to two of our State organizations to full membership and have recognized her as a co-laborer in a field full of interest, and one in my judgment to which she is well adapted. * * * Have not all parents who fail to give their daughters a profession or trade neglected one of the plainest requirements of life? Certainly. The world is full of misery on that account. I am sick of that can't and hypocracy that would prevent

women doing anything to earn her daily bread,
and then call it a dispensation of Providence when
she is left to support a family by spending her days
and nights over the needle. Let your daughters
enter the professions on anything they can earn a
livlihood at, and regard it as a dispensation of
Providence that he has in His wisdom given your
daughters brains enough to take a position in life
superior to that you, possibly, have ever been able
to fill."

In consequence of the liberal advantages offered
in dentistry to artists, scientists and students, we
to-day, as a profession, stand almost alone in the
realization of our imagination, the equal of any, the
superior of almost any other specialty in point of
success. But because of the fact we must not
permit the thought to become lethargic and assume
an attitude indifferent to our surroundings. For,
if we stop to sleep, others with that assiduity of
purpose and labor will excel us in their pursuits,
while we then will assume a position of mediocrity.
Man is too often deluded with false ideas of greatness; to cease to labor is a dead-lock to progress,
and to stop thinking, but another form or name for
imbecility. During the last two decades the improved methods, larger range and more exact style
of inquiry, and the assistance and hints which one
branch of study has given to others has produced
the most satisfactory results. The inquiries are

not yet complete; they seem on the contrary to have only commenced, and promise ultimately to satisfy all the useful purposes and legitimate curiosity of the many spectators.

Kind advice can be found in Dr. John S. Marshall's address recently delivered before a class of dental students.   He said:

"Knowledge proved and classified becomes science. The sciences underlie the intelligent practice of all the professions; consequently, to be educated for a profession means that you shall have knowledge of those sciences upon which it is based, and upon which it must depend for its intelligent practice. The dental student who commences his practice with the idea of obtaining his degree with just as little expenditure of time and energy as is possible under the rules of the institution with which he is connected, will make a dismal failure of both student and professional life. Justice will repay him in the same coin to the very last decimal, and in the same spirit with which they were meted out by him during his student days."

The doctor continued by saying: "To be successful in any profession in these times the individual must be well grounded in the fundamental sciences that underlie the superstructure of special professional knowledge; he must begin at the very foundation stones, and step by step go over every principle taught until he arrives at a correct under-standing of their application and their individual and mutual relationship and dependencies."

Thus, cherishing these well-worded sentiments, and knowing full well that you, too, agree with these worthy remarks, I began the history of dental prosthesis with the study of the subject at the very foundation stone, and step by step have gone over every principal taught.

To perfect yourself in this branch should be an upper thought of mind, and in order to accomplish this you will have no easy task, for, as Dr. Harris says, "Prosthetic dentistry constitutes by far the largest and most difficult part of dentistry, and thus makes it a distinct branch of the Art of Medi-cine and gives to it the power to add, as it does to health, comfort and the enjoyment of life."

As regard the benefit derived from an histori-cal review of dental prosthesis, such as I have earnestly endeavored to portray, I am of the same mind as Dr. Patrick, who once said: "I have been of the opinion that there is a growing desire in our profession to be more conversant with the ancient as well as the modern history of dentistry in its several departments—that there is a conviction that the literature of dentistry has been neglected. It is to be hoped that the time is not far distant when our profession everywhere will be convinced that the importance of becoming more intimately

acquainted with the researches and the views enter-
tained on the subject of dentistry by some of the
most intellectual men the world has ever produced.
It is to be hoped that the profession will see the
advantages and necessity of a dental encyclopaedia
or summary of dental knowledge; not a system,
but a work that would rescue valuable purposes
relative to dentistry that are now resting in com-
parative obscurity in the archives of dental society,
and that are now in a manner lost to the profession.   One great advantage the profession would
have in the possession of such a work would be,
that when a new theory is advanced, it could be
tested by comparing it with doctrines of a similar
nature advanced in former times.   Certainly every
age should profit by the experience of the preceding one; but without a record of history of what
has been accomplished, each investigator commences a new series of trials and wanders over the
same ground in research of truths which have long
ago been discovered.   The views of our predecessors may be justly regarded as beacon lights set
up to guide our footsteps from pitfall of error."*

One of the most gratifying evidences of the
progress of modern dentistry is the ever increasing
interest manifested in historical research and study
of the various branches of the science.   The profession is gradually comprehending that the proper

*Dental Review, vol. III., p. 436.

PLATE III.

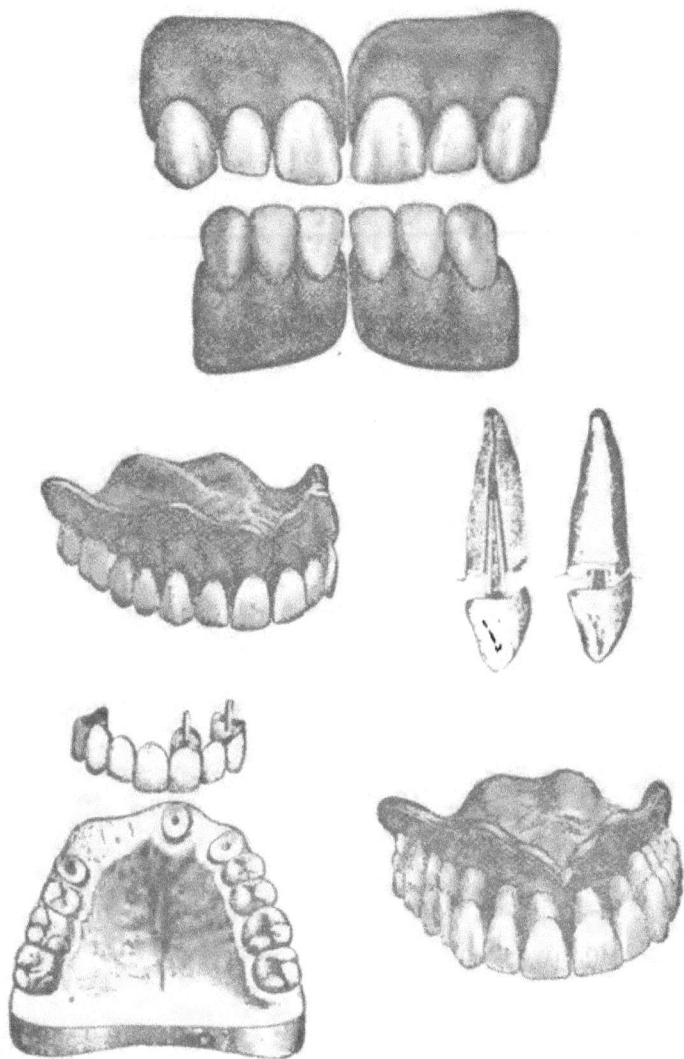

Specimens of Modern Dental Art.

way to learn lessons of wisdom for the uncertain future is, to give immediate attention to events of the past. All down the "long avenues of time" the voice of the departed are calling, giving us words of warning to avoid the errors which wrecked their successes and attempts. But how can we prevent a similar sad fate for our cherished plans and sail free from threatening perils if we heed not the advices of our forefathers, and remain ignorant of their accomplishments and the general status of our profession.

I confidently hope that those of you who have earnestly followed my remarks on the evolution of dental science will have enjoyed as much pleasure, and reaped a similar volume of information, as I have in the compilation of these historic facts; and I trust that my words shall have awakened in you the latent admiration for the profession, that you will from now on earnestly and persistently labor to "establish for our chosen profession a land-mark" among all sciences and vocations that future generations, instead of yielding but reluctant confidence, will then pour forth a full measure of respect and devotion.

APPENDIX.

JAPANESE.—His excellency, the Japanese Minister at Washington, explains that it was formerly an almost universal custom in Japan for women to signify their marriage by blackening their teeth.

There is a tradition that the practice was introduced by the wife of a famous Tycoon, who thus destroyed her beauty as a token of devotion to her lord and master. The tradition is silent as to the husband's sentiments at thus exchanging a beautiful bride for an ugly wife.

Another origin for this strange custom, which at first prevailed with both sexes, is supposed to have been the contract of the people of Japan with the natives of Annam, who chew the betal-nuts and leaves for the purpose of discoloring their teeth. This effect the Japanese were said to have imitated by other means. Whatever the origin, the custom has now almost disappeared except among old fashioned people and in remote parts of the Empire.*

---

*The author is obliged to Dr. J. Littlefield, who has charge of the S. S. White Dental Manfg. Co.'s exhibit at the Columbian Exposition, for the above information.

A grand display of Japanese dental instruments and workmanship can be seen at the Chicago branch of the S. S. White Dental Manufacturing Co.

PHŒNICIAN.—THIS ENGRAVING REPRESENTS THE ANCIENT DENTIST OF WHOM WE SPEAK ON PAGES 58, 59 AND 60.

A work has recently been issued in the Japanese characters on operative dentistry by the eminent Japanese dentist Dr. J. Watanabe of Tokio. The Japanese have also a dental journal published in native characters and is issued in Tokio.

AMERICAN.—The following are appended, to more fully complete the list of dental journals:

Southern Journal of Medicine and Dentistry. Published in 1853-1854.

Brown's Dental Advertiser, Cincinnati. Editor, J. M. Brown, 1854-1855.

New York Dental Journal and Reporter, New York City. Editor, F. H. Norton. 1858-1860.

Dental Science and Quarterly Art Journal, New York City. Editor, A. P. Merrill. 1875-1876.

Odontographic Journal, Rochester, N. Y. Editor, J. E. Line. 1879 ——.

The Dental Brief, St. Louis. Editor, F. T. Grimes. 1881-1882.

The Western Dental Journal, Kansas City. Editor, J. D. Patterson. 1886 ——.

The Southern Dental Journal and the Dental Luminary have consolidated under the name of the Southern Dental Journal and Luminary. Editor, H. H. Johnson. Published at Macon, Ga.

ERRATA.—Hippopotami instead of hippotami; line 4, p. 44.

Deutsch instead deutch; foot-note, p. 91.

Josiah instead Jossiah; line 21, p. 169.

Licentiate instead scentiate; line 20, p. 154.

Charge instead chare; line 21, p. 179.

One million instead 100,000; line 23, p. 194.

Maryland Dental College is defunct; line 9, p. 216.

Jefferis instead Jefferies; line 20, p. 218.

Chase instead Case; line 4, p. 230.

Luminary instead Summary; line 3, p. 231.

The Dental Record was edited by F. W. Leonard and was discontinued in 1882; line 28, p. 231.

Kimball instead Kimble; line 21, p. 233.

290 THE RISE, FALL AND REVIVAL

VALEDICTORY REMARKS.

"The dental profession has established and pro-
longed the reign of beauty; it has added to the
charms of social intercourse, and lent perfection to the accents of eloquence; it
has taken from old age its most un-
welcome feature, and length-
ened enjoyable human life
far beyond the limit of
the years when the
toothless and pur-
blind patriarch
might ex-
claim:—
I have no pleasure in them."

—DR. OLIVER WENDELL HOLMES.

NAME INDEX.

Smith, 34, 212, 213, 214,
219, 221, 227, 228.
Snowdin, 228.
Spalding, 230, 232.
Spencer, 117.
Spooner, 224, 225.
Stafford, 158.
Starr, 203.
Steel, 82, 100.
Stellwagen, 213, 214.
Stevens, 142.
Stevenson, 233.
Stockton, 192, 227.
Stolper, 158.
Storey, 232.
Sudduth, 220, 233.
Swain, 220.

T

Taft, 95, 142, 152, 212,
216, 227, 233.
Talbot, 95, 97.
Tandinier, 173.
Taylor, 212, 227.
Tees, 229.
Thailkill, 231.
Thais, 88.
Theon, 28.
Thomas, 202, 219, 222,
233.
Thompson, 153, 204,
220.
Tiberius, 92.
Timaeus, 147.
Toland, 228.
Tomes, 152.
Tover, 147.

Townsend, 212, 213.
Trenor, 224.
Truesdell, 232.
Truman, 153, 213, 217,
233, 277.
Tuller, 147.
Turnefort, 66.
Tyson, 229.

U

Ulrich, 216.
Underwood, 232.
Urban V., Pope, 94.

V

Van Antwerp, 212.
Van Der Belen, 144.
Van Der Maessen, 144.
Valentine, 144.
Van Marter, 60, 95, 104,
105, 106.
Van Rhyn, 103.
Vesalius, 123, 144.
Virgil, 89, 95.

W

Wagoner, 230.
Walker, 71, 76, 90, 222.
Walkins, 221.
Walton, 160.
Wardle, 214.
Warren, 39, 233.
Washburn, 28, 30.
Washington, 166, 177
179, 180, 188, 192.
Watling, 216.
Watson, 166, 167.

Watt, 157, 231.
Webb, 202.
Webster, 200.
Weden, 229.
Weeks, 220.
Weiss, 162.
Welchens, 230.
Welch, 186, 231.
Wello, 107.
Weston, 202.
Westcoat, 211, 212, 213.
Wetherbee, 216.
White, 193, 194, 227, 229.
Whitelock, 169.
Whitmore, 219.
Whistlar, 222.
Wildman, 193, 213, 229.

Wilkerson, 231.
William, King, 114.
Williams, 202, 230.
Willard, 232.
Wilson, 220, 222.
Winder, 211, 212, 216
Wood, 229.
Woodward, 214.
Woofendale, 147, 149
    164, 165, 200, 223.
Wright, 260.

Y

Young, 230.

Z

Ziegler, 229.

SUBJECT INDEX.

A

Dental Journals, continued:

312   THE RISE, FALL AND REVIVAL

T

V

W

www.ingramcontent.com/pod-product-compliance
Lightning Source LLC
Chambersburg PA
CBHW021503210326
41599CB00012B/1112